Doing Statistical Analysis

Doing Statistical Analysis looks at three kinds of statistical research questions – descriptive, associational, and inferential – and shows students how to conduct statistical analyses and interpret the results. Keeping equations to a minimum, it uses a conversational style and relatable examples such as football, COVID-19, and tourism, to aid understanding. Each chapter contains practice exercises, and a section showing students how to reproduce the statistical results in the book using Stata and SPSS. Digital supplements consist of data sets in Stata, SPSS, and Excel, and a test bank for instructors. Its accessible approach means this is the ideal textbook for undergraduate students across the social and behavioral sciences needing to build their confidence with statistical analysis.

Christer Thrane holds a PhD in Sociology and is Professor at Inland Norway University of Applied Science – Inland School of Business and Social Sciences. His research interests include quantitative modeling studies in leisure, sports, and tourism. He has 25 years of experience teaching quantitative research methods, and he has written several textbooks on this topic.

Doing Statistical Analysis

A Student's Guide to Quantitative Research

Christer Thrane

Routledge
Taylor & Francis Group

LONDON AND NEW YORK

Cover image: © watchara_tongnoi / Getty Images

First published 2023
by Routledge
4 Park Square, Milton Park, Abingdon, Oxon OX14 4RN

and by Routledge
605 Third Avenue, New York, NY 10158

Routledge is an imprint of the Taylor & Francis Group, an informa business

British Library Cataloguing-in-Publication Data
A catalogue record for this book is available from the British Library

Library of Congress Cataloging-in-Publication Data
Names: Thrane, Christer, author.
Title: Doing statistical analysis : a student's guide to quantitative research /
Christer Thrane.
Description: Abingdon, Oxon ; New York, NY : Routledge, 2022. |
Includes bibliographical references and index. |
Identifiers: LCCN 2022007878 (print) | LCCN 2022007879 (ebook) |
ISBN 9781032180304 (hardback) | ISBN 9781032171326 (paperback) |
ISBN 9781003252559 (ebook)
Subjects: LCSH: Social sciences—Research—Statistical methods. |
Research—Statistical methods.
Classification: LCC HA29 .T547 2022 (print) | LCC HA29 (ebook) | DDC
519.5—dc23/eng/20220224
LC record available at https://lccn.loc.gov/2022007878
LC ebook record available at https://lccn.loc.gov/2022007879

ISBN: 978-1-032-18030-4 (hbk)
ISBN: 978-1-032-17132-6 (pbk)
ISBN: 978-1-003-25255-9 (ebk)

DOI: 10.4324/9781003252559

Typeset in Bembo
by codeMantra

Access the Support Material: www.routledge.com/9781032171326

Contents

Acknowledgments

Thanks to Natalie Tomlinson for the opportunity to publish this textbook on Routledge. Thanks also to the reviewers for valuable suggestions to the book project at the outset, and to Henning Bueie for initially helping me with some of the book's figures. Per Arne Tufte deserves a big thank-you for comments on the complete book draft. All remaining errors and ambiguities are, of course, my responsibility.

I recently published a textbook on regression analysis on Routledge (Thrane, 2020). Some of the material in the present book, especially in Chapters 4 and 6, draw heavily from my regression book. I also published two Norwegian textbooks on quantitative research methods/statistics in 2018 and 2020. Some of the material in these books have been adapted and have found its way into the present volume. A big thank-you to all those contributing to my previous Norwegian textbooks.

Lillehammer, Norway, November 2021
Christer Thrane

Acknowledgement

1 What Is Statistical Analysis from a Research Perspective?

1.1 A Statistical Association: COVID-19 Spread and Residential Property Prices

Oslo is the capital of Norway. Like many such capitals, Oslo is made up of several city districts. Table 1.1 contains some quantitative information on three of these districts. I have named the districts A, B, and C because Norwegian names are unfamiliar and more than a mouthful to pronounce for non-Norwegians.

We find the three districts' COVID-19 infection rates in the right column of the table. These rates express the number of people infected per 100,000 persons. We note that district B has the largest infection rate, followed by district C, and district A. The table also provides information on how costly it is on average to buy a residential property in these three districts. We note that properties in district C are most pricey, followed by the properties in district A and district B.

Statistical analysis in research is often about associations or relationships.[1] We might ask if there is some kind of association between COVID-19 rates and the average property prices for the six numbers in Table 1.1. In this regard, we note that district B has the least expensive properties and the largest COVID-19 infection rate. In contrast, districts A and C, both pricier at about the same level, have lower infection rates. We thus note the contour of a systematic association. Yet, since we base our visual statistical analysis (for lack of a better term) on only three districts, we cannot be sure about this.

As it happens, a total of 15 districts make up the city of Oslo. Table 1.2 presents the complete quantitative information about all these districts.

Table 1.1 COVID-19 infection rates and average residential property prices per square-meter in thousands of Norwegian Crowns (NOK) for three of the city districts in Oslo, Norway's capital.

Oslo districts	Average property price per square-meter	COVID-19 infection rate
District A	79.470	231.20
District B	43.210	448.00
District C	84.772	310.70

Note. The COVID-19 infection rates refer to November 2020, as reported in the Norwegian newspaper *Aftenposten*. The residential property prices per square meter in NOK (1,000) pertain to 2019.

DOI: 10.4324/9781003252559-1

Table 1.2 COVID-19 infection rates and average residential property prices per square-meter in thousands of Norwegian Crowns (NOK) for the 15 city districts in Oslo, Norway's capital.

Oslo districts	Average property price per square-meter	COVID-19 infection rate
District A	79.470	231.20
District B	43.210	448.00
District C	84.772	310.70
District D	69.166	185.40
District E	78.535	112.80
District F	90.332	246.30
District G	76.616	351.10
District H	63.190	217.30
District I	58.627	261.80
District J	79.139	298.00
District K	52.754	455.80
District L	45.389	441.20
District M	50.337	458.40
District N	59.782	511.60
District O	83.082	224.00

Table 1.2 contains more information than Table 1.1, but here we face a new problem. By simply looking at the 30 numbers it is hard to visualize any potential association between property prices and infection rates. Yet, if we put the property prices on the x-axis and the infection rates on the y-axis of a scatterplot, as in Figure 1.1, the 30 numbers transform into a cloud of districts with a systematic pattern. The straight line in the figure captures the trend in the pattern, which we call a negative association: Higher residential property prices seem to go hand in hand with, or are associated with, lower COVID-19 infection rates. Conversely, lower prices appear to suggest higher rates. Why do we see this pattern? The association in itself does not tell us why, but we may speculate. The high prices in some districts might reflect large residences and spacious surroundings, that is, an environment that makes it harder for the COVID-19 virus to spread. Alternatively or additionally, the association may reflect differences between the lifestyles of people living in pricy and those living in less pricy districts. Whatever the cause might be, the fundamental point is that a statistical analysis may find an association between two phenomena but generally tells us little about why we observe the association in question. To explain such an association causally – that is, to answer why we observe what we observe – we also need a theory, reason, or mechanism to guide us.

Figure 1.1 shows a systematic statistical pattern – that is, a negative statistical association – between property prices and COVID-19 infection rates. Put differently, the figure answers the associational research question of how property prices are associated with COVID-19 infection rates. Such a research question is one of the three main types of statistical questions we are about to take on in this book. The remaining two are descriptive and inferential questions. We return to these three types of questions in Section 1.3. Before that, however, we should briefly mention what statistical analysis in research contexts is all about and answer the question of why and how we examine this in the present book.

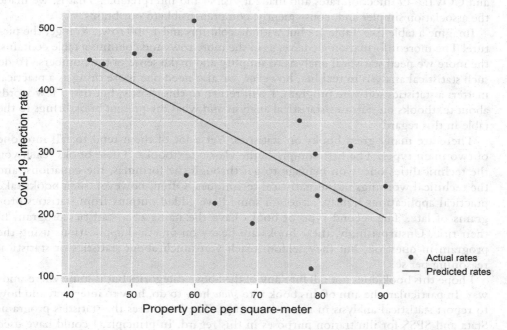

Figure 1.1 Scatterplot of residential property prices and COVID-19 infection rates for 15 city districts in Oslo, with trend line.

1.2 Why Do Statistical Analysis in Research? The Book's Purpose and Pedagogical Approach

Using a scatterplot and a trend line on the numbers in Table 1.2, as in Figure 1.1, is to do a statistical analysis. Why do this? A simple answer is that we do statistical analysis in research to increase our knowledge on some topic. I repeat: Statistical analysis is about crunching numbers to become wiser than we were before. In this regard, the present book is about to take on three intertwined aspects of statistical analysis from a research perspective:

1 Doing statistical analysis
2 Interpreting the results of statistical analysis
3 Presenting the results of statistical analysis to readers/audiences

The reason for doing statistical analysis also has a subtler answer: We do it to simplify large amounts of quantitative information so that such information becomes comprehensible in our minds. This might sound odd on the face of it because most people think of statistical analysis as anything but simple. To wrap your head around it, look at Table 1.2 again! It is very hard to discover, or to make any sense of, the association between the two phenomena simply by looking at the table. There is too much information. However, by using a scatterplot and a trend line – that is, by doing statistical analysis – we unraveled the hidden and negative association between property prices and COVID-19 infection rates and made it visible and interpretable. That is, we made the association simpler and, consequently, comprehensible to our brains.

Imagine a table like Table 1.2 but with 30 columns and 1,500 rows. You get the picture! The more information we have, as in the more rows and columns a table contains, the more we need statistical analysis to simplify and make sense of the numbers. To do such statistical analysis in real life, however, we also need one more thing as a practical matter: a statistics software program. I will return to this shortly, but first a few words about textbooks on statistics/statistical analysis and what the present book brings to the table in this regard.

There are many great books on statistics. Yet a lot of them tend to fall into one of two main types. The first comprises the classic textbooks. These books teach you the technicalities once you manage to get through the formulas, the equations, and the technical workings of the statistical techniques. Often, however, such books take practical applications lightly, although some have added outputs from statistics programs of late. The second type of books have the name of a statistics program in their title. Unsurprisingly, these books are heavy on practical applications using the program in question, but they seldom teach you much about statistics or statistical reasoning per se.

I hope this book does not fall into any of the above categories but is somewhere midway. In particular, the aim of this book is to *show* how to do, how to interpret, and how to report statistical analysis in research settings. The book uses the statistics programs Stata and SPSS for illustration purposes in this regard. In principle, I could have used a plethora of statistics programs (e.g., Statistica, R, SAS, or Gauss) to do the calculations. The reasons for preferring Stata and SPSS are twofold: First, both programs are popular and easy to use. Second, I have used both of them in my teaching during the last 25 years. That said, you could download the data sets accompanying this book, do the analyses in the statistics program of your choosing, and still learn all that the book has to offer.

A second feature that I hope distinguishes this book from many others has to do with the background or context of the statistical questions. That is, many books are often rather silent on the process with which such questions are being asked and answered. One of my aims in this book is to get the research context back into statistical analysis. In most of the book's examples, thus, the context or background is some research finding, some idea, or some everyday fact. Experience tells me that context aids learning of statistical analysis, whereas context-free demonstration of statistical techniques undermines it. That said, since too much context hinders forward thrust, which I also believe in, there is a balance in play. In this regard, I use notes for supplemental information that might have stalled forward thrust or interrupted the narrative if placed in the main text.

The book differs from others in other respects as well. I introduced the main event of most of the statistical analysis in research – that is, the study of associations – on the book's first page. In contrast, most books tend to do this in the second half or later. That is, elsewhere you have to consume many unnecessary appetizers before getting to the main course of the meal.[2] Furthermore, the book pays more attention to the presentation of the results of statistical analysis to readers and audiences than what seems to be the norm in other books on statistical analysis.

Many introductory texts on statistics are technically thornier than they need to be in mathematical terms. This scares off many readers. An issue separating this book from others is thus the die-hard principle of keeping matters as simple and down-to-earth as possible. To hold on to this keep-it-simple approach, I follow the example-driven and applied path. The book also focuses on what the student should know when the data appear on their computer and their objective is to analyze them. Many comprehensive textbooks on statistics (which we also need) concern topics not relevant in this regard, and I skip them for the most part. However, I will point to such books when needed. Finally, I aim to make statistical analysis relevant and interesting for many types of students – not just students following specific subjects. To achieve this goal, I use examples backed up with easily recognizable data.

It might be redundant to spell out the intended audience for this book against the background just provided. Yet I will do so at the risk of over-explaining. The student who will benefit most from the following pages is one who needs to do statistical analysis and who also finds traditional books on statistics too abstract, too technical, too long, too irrelevant, or too boring. I realize I lay my head on the block here!

1.3 Three Types of Statistical Research Questions: Descriptive, Associational, and Inferential

The book is organized around three types of statistical research questions. The descriptive questions deal with the description of large amounts of information in a summary manner. Consider the 15 COVID-19 infections rates in Table 1.2. A way to describe these rates more generally is to come up with an answer to the question of what the *typical* infection rate is among all the 15 districts. I return to this in Section 1.5 and Chapter 2. The second type of statistical research question is of the associative kind, such as the one between property prices and COVID-19 spread in Oslo. I have more to say on such questions in Section 1.5 and Chapters 3, 4, and 6. The third type of statistical question is the inferential one. The information in Table 1.2 concerns the 15 districts of Oslo. May we use the statistical association between property prices and COVID-19 spread in Oslo to say something about similar relationships in other capital cities? This is an inferential question, and I will say more on this topic in Section 1.5 and Chapters 5 and 6. Before taking on Chapters 2 to 6, however, we first need to cover some groundwork. This is the topic of the remainder of Chapter 1.

1.4 Some Key Concepts You Really Should Understand

Bear with me! There are some key concepts you must master to become a skilled statistical analyst. On the brighter side, however, you already met many of them in Section 1.1. Yet there I eschewed statistics lingo for the benefit of everyday language. The most basic concepts include:

1.4.1 Data

I mentioned the terms "quantitative information" and "numbers" in Sections 1.1 and 1.2 by referring to the contents of Tables 1.1 and 1.2. The shorter term in this regard is simply *data*. That is, Table 1.2 contains the data or raw data for the 15 city districts of Oslo. We might also say that Table 1.2 is a data set or a data matrix. Generally, we do statistical analysis of, or on, our data or data set.

1.4.2 Variables

We speak of phenomena, features, traits – and some of us, informally, of things – in everyday life. In statistics lingo, however, there are no such "things." That is, in statistics lingo we write and speak of *variables*. What is a variable? A variable is something that varies. (No surprise there!) Remember the COVID-19 data in Table 1.2. The infection rates varied among the 15 districts; some districts had smaller rates and some had larger rates. The take-home message is that statistical analysis is about the analysis of variables. Variables appear as columns in data sets, and a variable's name most often appears on the top of a column, as in Table 1.2.

1.4.3 Units

A variable is something that varies among a set of units. Other names for such units are observations or cases.[3] The units, observations, or cases in our COVID-19 data are thus the 15 districts of Oslo. More generally, units are often persons, firms, counties, or countries in the social and behavioral sciences. Yet they could be anything: trans- actions, stocks, products (e.g., cars, houses, wines), or services (e.g., meals, hotel stays). When people answering a survey questionnaire are the units, we typically call them respondents or subjects. Formally, we have variable information on a set of units. The units make up the rows in data sets, as in one district's variable information for each row in Table 1.2.

1.4.4 Variable Value

Variables take on different values. The values for the COVID-19 rate variable in Table 1.2 are between about 113 and 512. Similarly, the average residential property prices per square meter all lie in the range between 43,000 and 90,000 NOK. Other types of variables, which I return to in Chapter 2, might take on many more or fewer potential values.

1.4.5 Variables' Measurement Levels

The value of a COVID-19 infection rate is a number, and the same goes for an average property price. We call variables having numbers as their values for continuous variables or numerical variables (to simplify a bit). Alternatively, we might claim that such vari- ables are on the continuous or numerical measurement level. Other variables are more categorical by nature, and I introduce such variables in Sections 2.3 and 2.4. The gender variable is a case in point having two values: female or male. A variable's measurement level has consequences for what kind of descriptive questions being relevant (cf. chapter 2) and for how to address associative research questions (cf. chapters 3, 4, and 6).

1.4.6 Independent and Dependent Variable

The tacit assumption behind the association between property prices and COVID-19 infection rates was that the former somehow affected the latter: property prices → infection rates.[4] "Assumption" is the keyword here. When analyzing statistical associ- ations, we tend to assume that variation in one variable is responsible for variation in another variable. Yet we do not like using the expression "cause variation in another variable," and I explain why in Sections 3.6 and 3.7. When examining a statistical association, however, we must always tell the statistics program about the assumed causal direction of our association. We use the terms independent variable and dependent variable in this regard. The independent variable is the one we assume brings about variation in some other variable. The dependent variable is the one being affected by an independent variable. It is common practice to denote the independent variable as x and the dependent variable as y. Figure 1.2 sums up.

In the data for a bachelor or master's thesis, x and y might be anything that varies among a set of units. More generally, our imagination is the only boundary for what might go into the statistical analysis of an association between x and y in such a research setting. For convenience, I mostly use x or x-variable and y or y-variable in this book rather than longer independent and dependent variables.

1.4.7 The Place for Statistical Analysis in the Quantitative Research Process

Many textbooks have defined the quantitative research process as a series of phases or steps (e.g., Bryman, 2016). The first phase covers the start of the project, as in the first loose ideas to the final research question (RQ). The RQ is often an expectation of an association between an independent and a dependent variable, x and y. We often call this associational expectation a hypothesis. We will learn to test such hypotheses formally in Chapter 5.

The second phase is about obtaining the data to answer the RQ. We generally have two options in this regard: We may collect new data ourselves, as in making a survey questionnaire or by compiling the relevant numbers from, say, the Internet. Alternatively, we may use already available data collected by others. This phase typically also includes the cleaning and preparation of data to get them into their final form and ready for analysis. In this book, I have gathered and prepared the data for you. (Lucky you!) I will take on some important aspects of data preparation in Chapter 6, though.

The third phase is about doing and interpreting statistical analysis: descriptive analysis (Chapter 2), associational analysis (Chapters 3, 4, and 6), and inferential analysis (Chapters 5 and 6). This book has, unsurprisingly, most to offer on this phase.[5]

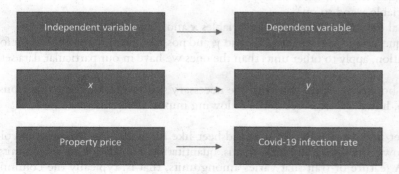

Figure 1.2 Independent and dependent variable. The arrows show the assumed causal directions of the statistical associations, that is, from independent to dependent variable or from x to y.

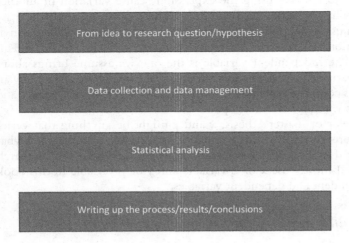

Figure 1.3 The quantitative research process as a series of phases.

Fourth and finally, we write up the research process and present our results and conclusions in a thesis or a research paper – or in a PowerPoint presentation. Yet I advise you to start the writing-and-presentation process earlier than this. I will say more on writing up and presenting statistical results in Chapter 6. Figure 1.3 sums up the quantitative research process from beginning to end.

1.5 Chapter Summary, Key Learning Points, and the Organization of the Rest of the Book

Why do statistical analysis in research? We do it to reduce the complexity of large amounts of quantitative information to become more knowledgeable about some topic. The information we analyze we call our data, which are typically stored in spreadsheet-like matrixes. Statistical analysis in research spins around three intertwined types of questions:

- Descriptive questions: What is the typical value of variable x, and what is the typical value of variable y and so on?[6]
- Associational questions: How are the variables x and y associated, if at all?
- Inferential questions: Do the values of x and y, and possibly the numeric expression for their association, apply to other units than the ones we have in our particular data set?

The aim of this book is to equip you with the necessary skills to tackle these questions. Before doing so, however, let us recap the following important terms:

- Data/data set/raw data: Typically, a spreadsheet-like file with large amounts of columns and rows containing numbers, that is, quantitative information on many units.
- Variables: A feature or trait that varies among units, that is, typically the columns in data sets.
- Units/observations/cases: The entities for which variables vary. Typically, persons, firms, counties, or countries in the social sciences, but they could be anything. Most often, the units make up the rows in data sets.

- Variable value: The possible values a variable may take on for a unit.
- Independent variable/x: The variable we assume affects another variable.
- Dependent variable/y: The variable we assume is affected by an independent variable/x.

I recommend reading the upcoming chapters in the order in which they appear. That said, if you have a good grip on descriptive statistics, you might jump directly to Chapter 3. The remainder of the book is organized as follows:

Chapter 2 dives into descriptive questions: What is typical for a variable? We start on a small scale with the COVID-19 data and with a focus on what characterizes the infection rate variable. Next, we turn to the analysis of variables in larger data sets. These include variables characterizing Christmas beers, variables characterizing Norwegian soccer players, and variables describing college students' health behaviors. Along the way, I explore the different types of variables in quantitative research by addressing their different measurement levels. Chapter 2 is, in short, about doing and presenting descriptive statistics. It is also a stepping-stone for the chapters to come.

Chapter 3 takes on the main event of most statistical analysis done in research: the associational statistical questions. I will show that we typically face three types of associational or relational research questions: questions regarding differences in proportions, questions regarding differences in averages, and correlational questions. Please relax; these terms will be explained when we get there! To illustrate the statistical techniques corresponding with the three types of research questions, I will continue using the variables and data sets introduced in Chapter 2. Towards the end of Chapter 3, in Sections 3.6 and 3.7, I discuss the limitations of associational analysis involving only two variables. These two sections foreshadow Chapter 4.

Chapter 4 first spins around one vital question: How can we be sure x is associated with y when we analyze observational data? That is, how do we know it is the variation in x and not the variation in some other variable, say z, that brings about the variation in y? Multiple regression answers this (causal) question and many more in the behavioral and social sciences when we have no access to experimental data. Later in Chapter 4, I address more complex multiple regression scenarios. I also introduce some new data sets for illustration purposes in this chapter.

Chapter 5 tackles the inferential questions. In research, we typically search for conclusions stretching beyond our specific data; we aim for results that should pertain to more people, more situations, more places, and more cultures. If our data are representative of some larger entity, we might justify generalizing our results in this way. In contrast, if our data are not representative of such an entity, such an inference appears dubious. Inferential statistics is about the conditions we need to make trustworthy (valid) generalizations from our data. Often, but not always, this involves the analysis of a random sample from some well-defined population. (Yes, I will explain these concepts as well when we get there!)

When you get to Chapter 6, you will have gotten a firm grasp on doing statistical analyses and on how to interpret the results of such analysis in a research context. Yet, things seldom go as smooth as presented in this book in real-life research. Much of the first part of Chapter 6 thus solves a number of problems that tend to come up in actual research. The last part of Chapter 6 deals with how to present and communicate to readers and audiences the results of the statistical analyses covered in Chapters 3–5.

Notes

1 I use the term statistical analysis throughout the book for simplicity, rather than the more appropriate statistical *data* analysis. I use the terms association and relationship interchangeably, but mostly the former.

2 I mean "unnecessary" in the sense that many of the topics typically placed in the first part of statistics books (e.g., descriptive statistics, inferential statistics, operationalization, variables' measurement levels, etc.) are by no means necessary to get a firm grip of the concept of a statistical association. See White and Gorard (2017) for a similar view.

3 We may use units and observations interchangeably as long as the unit occupies one row in the data. When the same unit occupies two or more rows in the data, as in having several observations for the same unit, we should probably distinguish between units and observations.

4 In principle, however, the causal arrow could go the other way round: property prices ← infection rates.

5 A wishful side effect on my part: After having read the book, I hope what you have learned will make you see more possibilities than you did before regarding new ideas for quantitative research questions/hypotheses.

6 This is a simplification. Descriptive statistical questions also concern variables' distribution and spread; more on this in Section 2.6.

2 Descriptive Research Questions

2.1 Introduction and Chapter Overview

This chapter deals with descriptive research questions or descriptive statistics. Descriptive statistics is the starting point of all statistical analysis in research no matter what the end goal might be. The key question of descriptive statistics is very often this: What is typical for a variable?

Section 2.2 introduces Stata and SPSS using the COVID-19 data. Here, I also bring up the advice to work commando-style with statistics (more on this in Section 2.9). Then I proceed with the analysis of a data set containing 75 Christmas beers. Section 2.2 elaborates on the topic of what is typical for a variable, which in statistics lingo translates into three measures of central tendency: mean, median, and mode.

Section 2.3 takes on variables' measurement levels. I continue using the data on Christmas beers, but I also introduce two more data sets – one on the attributes of Norwegian soccer players, and one on college students' health behaviors – to shed more light on variables' different measurement levels.

Section 2.4 is about ordinal variables. I treat ordinal variables as a special case regarding variables' measurement levels for presentational ease. This section explains why. Sections 2.2 through 2.4 present the results of descriptive statistical analyses mainly as tables. Section 2.5 presents the same kind of results using graphs. Section 2.6 concerns the variation or spread of continuous variables and not their central tendency (which is the focus of Section 2.2 and partly of Section 2.3 also). Section 2.6 is also the stepping-stone for Chapter 5. Section 2.7 sets the stage for Chapter 3, and Section 2.8 summarizes the chapter and lists the key learning points.

Section 2.9 reintroduces commando-style statistical analysis by means of do-files in Stata and syntax-files in SPSS to obtain reproducible results, whereas Section 2.10 provides exercises with solutions.

Note! Throughout Chapter 2, my comments regarding the various descriptive results pertain only to what happens *within* the data. I do not refer to what might happen (or not) outside of the data. The latter inference topic is for Chapter 5.

2.2 What is Typical? Three Measures of Central Tendency: Mean, Median, and Mode

The combination of observing and analyzing the data simultaneously is in my experience great for learning. I thus repeat the COVID-19 data from Section 1.1 in Table 2.1 below. The data set is called res_prop_price_COVID-19.[1] You find these data and most of the other data sets used in the book on the book's website, ready for download.

DOI: 10.4324/9781003252559-2

Table 2.1 COVID-19 infection rates and average residential property prices per square-meter in thousands of Norwegian Crowns (NOK) for the 15 city districts in Oslo, Norway's capital.

Oslo districts	Average property price per square-meter	COVID-19 infection rate
District A	79.470	231.20
District B	43.210	448.00
District C	84.772	310.70
District D	69.166	185.40
District E	78.535	112.80
District F	90.332	246.30
District G	76.616	351.10
District H	63.190	217.30
District I	58.627	261.80
District J	79.139	298.00
District K	52.754	455.80
District L	45.389	441.20
District M	50.337	458.40
District N	59.782	511.60
District O	83.082	224.00

2.2.1 Introduction to Stata

When opening the Stata statistics program, something similar to Stata-output 2.1 will pop up on your computer:

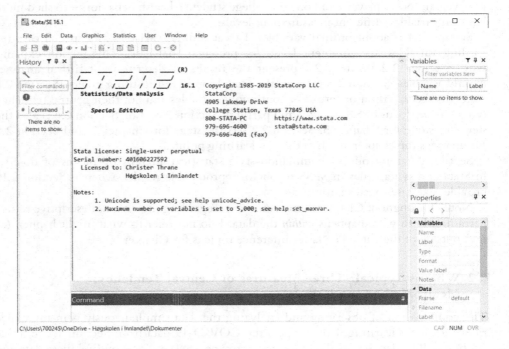

Stata-output 2.1 Stata's interface window.

The Command-window at the bottom is where you first type your statistical commands before pressing Enter (i.e., Carriage Return) to execute. The results then appear in the large Results-window in the middle of the screen. The names of the variables in the data appear in the upper right corner of the output – the Variables-window – once you have loaded the data into Stata. Stata-output 2.2 shows this for the COVID-19 data, where we have three variables: district, infect_rate, and price_sq_m.[2]

Stata-output 2.2 Stata's interface window, including the COVID-19 data.

What is the typical infection rate among the 15 city districts in Table 2.1? More generally, what is the typical value of variable *y* or variable *x*? In statistics lingo, we translate these questions into finding out about a variable's *central tendency*. The three most frequently used measures of central tendency are the mean, the median, and the mode. I address them in turn below, that is, in the order of their popularity.

2.2.2 The Mean

We find the mean or average of the COVID-19 infection rate variable for the 15 city districts in Table 2.1 by adding the individual districts' rates and dividing the sum by 15: $(231.20 + 448.00 + 310.70 + ... + 224.00)/15 \approx 316.90$.[3] Or, since we have a statistics program at our disposal, we make Stata do this calculation for us by typing

```
sum infect_rate
```

in the Command-window and pressing Enter. The output in the Results-window in Stata-output 2.3 then appears. The output from most other statistics programs has a similar visual appearance, and this is the case in general throughout the chapter. We see that Stata finds the Mean or average to be 316.9067. We note that Obs is short for observations, and that the data contain 15 units.

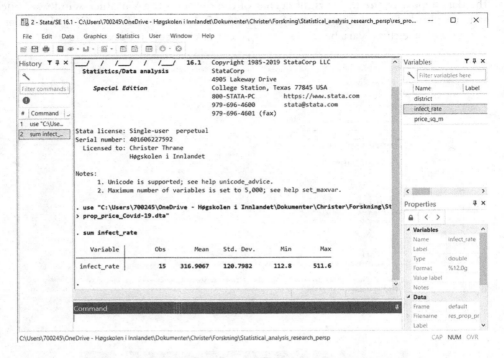

Stata-output 2.3 The mean or average of the variable infect_rate in the COVID-19 data.

The last thing to note in the output, for now, is the Min and Max columns. Unsurprisingly, they tell us about the lowest and highest infection rates in the data for the 15 districts. One substantive question remains unanswered: Is the mean COVID-19 infection rate of 317 a high or a low rate? A meaningful answer to this question requires information from outside of the data. As Stigler has eloquently put it (Stigler, 2016, p. 63): 'A measurement without context is just a number.' In the present case, this context could be the mean rate for another city, say London, or a threshold rate for some agreed-upon dangerous infection rate level.

2.2.3 Introduction to SPSS

When opening the COVID-19 data in SPSS, you will notice that the data appear on your computer screen much like in SPSS-output 2.1. There might be small differences in the setup between different computers and different versions of the program, however.

In contrast to Stata's typing-short-commands structure, SPSS uses a point-and-click or dialog box routine as the default way of generating statistical results. (Stata also has such a dialog box routine. In my experience, however, most users find the command-approach in Stata more intuitive.) To get the analogous results as in Stata-output 2.3, you click on Analyze → Descriptive Statistics → Descriptives. Once there, you drag the variable infect_rate from the box on the left and over to the empty box on

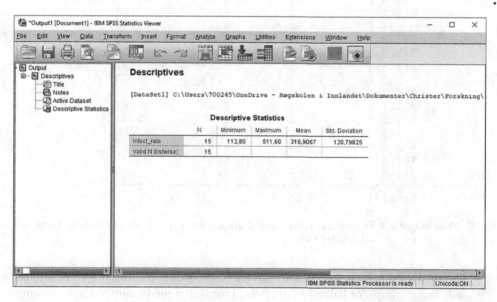

SPSS-output 2.1 The interface window of SPSS, including the COVID-19 data.

the right. Then you click on OK, and the output in SPSS-output 2.2 appears. The only difference between Stata and SPSS in this case is that the latter uses N to refer to the number of units and not Obs.

Descriptives

[DataSet1] C:\Users\700245\OneDrive - Høgskolen i Innlandet\Dokumenter\Christer\Forskning\

Descriptive Statistics

	N	Minimum	Maximum	Mean	Std. Deviation
infect_rate	15	112,80	511,60	316,9067	120,79825
Valid N (listwise)	15				

SPSS-output 2.2 The mean or average of the variable `infect_rate` in the COVID-19 data.

2.2.4 Some Formalities about the Mean

More generally, and replacing the infection rate variable with y, we have that

$$\text{mean of } y = \frac{\sum y}{n},$$

where $\sum$ is the summation sign for the individual values of y (i.e., COVID-rates) and n is the number of units (i.e., districts) in the data. To some, the use of such formulas serves as a convenient and quick notational shorthand. Since I do not belong in that category of people (by a long shot, I might add), I will eschew formulas and equations to the greatest possible extent in the pages to come.

2.2.5 The Median

If the mean is the most frequently used measure of a variable's central tendency, the median is the second-most popular choice. In Stata, we might find the median of the infection rate variable by typing

```
tab infect_rate
```

in the Command-window and clicking on Enter in the usual manner. I will skip the entire Stata interface from now on and report only what comes up in the Results-window; cf. Stata-output 2.4.[4]

```
. tab infect_rate
```

infect_rate	Freq.	Percent	Cum.
112.8	1	6.67	6.67
185.4	1	6.67	13.33
217.3	1	6.67	20.00
224	1	6.67	26.67
231.2	1	6.67	33.33
246.3	1	6.67	40.00
261.8	1	6.67	46.67
298	1	6.67	53.33
310.7	1	6.67	60.00
351.1	1	6.67	66.67
441.2	1	6.67	73.33
448	1	6.67	80.00
455.8	1	6.67	86.67
458.4	1	6.67	93.33
511.6	1	6.67	100.00
Total	15	100.00	

Stata-output 2.4 Frequency table (distribution) for the variable `infect_rate` in the COVID-19 data.

The median for the infection rate variable is the value splitting the 15 rates into two equally large groups (or halves) of districts. That is, the median infection rate comprises the middle row shaded in grey in the frequency table in Stata-output 2.4. Seven districts have lower rates than the median of 298, and seven have higher rates than the median.

Another way to get the median is to find the infection rate accounting for 50 percent of districts with the lowest infection rates. In this respect, the rate of 261.80 accounts for 46.67 percent of the districts; see the column Cum. in Stata-output 2.4. This is not enough. In contrast, the rate of 298 makes up for 53.33 percent of the districts, which is enough. The rate of 298 is in other words the median – which we already knew.

Any assessment of a median of 298 as high, low, or somewhere in between is contingent on information outside the data, as in the case of a mean of 317. The mean and the median are quite similar in this case. In others, they are not. We return to such examples later in the chapter.

To get a similar frequency table in SPSS, we click Analyze → Descriptive Statistics → Frequencies. Once there, we drag the variable infect_rate from the box on the left over to the empty box on the right. Then we click on OK, and the output in SPSS-output 2.3 appears.

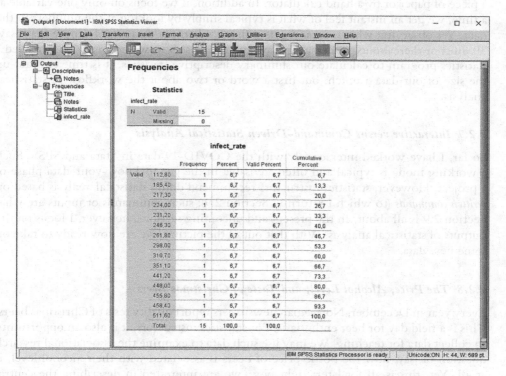

SPSS-output 2.3 Frequency table (distribution) for the variable infect_rate in the COVID-19 data.

2.2.6 The Mode

The mode is the third measure of a variable's central tendency or typicalness. The mode is also the measure of central tendency that applies to all sorts of variables; a feature I will get back to in Section 2.3. The mode is simply the *most frequent* value in a frequency distribution. Because no two districts have exact equal infection rates in our data, there

Table 2.2 Descriptive statistics for the infection rate variable in the COVID-19 data.

Variable	N =	Mean	Median	Mode
Infection rate	15	317	298	–

is no mode. We return to other cases in which the mode is a more meaningful summary statistic later in the chapter.

Summing up, the mean infection rate is 317, the median infection rate is 298, and the mode is not defined. If we were to produce a summary table of our findings in a thesis or a research paper, it would look something like Table 2.2.[5] We do not normally copy-paste the output from statistics programs directly into such publications, or in a PowerPoint for that matter, in most circumstances; see Section 6.8 for more on this.

The numbers in Table 2.2 refer to only 15 units, making it easy to calculate them on a piece of paper or by a hand calculator. In addition, if we focus on only one variable at a time, we get an instant feel of what is typical simply by looking at the numbers in the data. Yet observing what is typical gets difficult when the data refer to more than, say, 30 units or thereabouts. When analyzing larger data sets, as we normally do, we need a statistics program to calculate our summary descriptive statistics. It is time to scale-up the size of our data a notch, but first a word or two about the workflow of statistical analysis.

2.2.7 Interactive versus Command-Driven Statistical Analysis

So far, I have worked interactively with the COVID-19 data in Stata and SPSS. Such a working mode is typical and often advisable in the get-to-know-your-data phase of a project. However, statistics instructors recommend doing statistical analysis based on *written commands* (to which I return in Section 2.9); such commands or inputs are what Section 2.9 is all about. In the present and upcoming sections, however, I focus on the outputs of statistical analysis. With that out of the picture, we are now ready to take on some new data.

2.2.8 The Price, Alcohol Level, and Taste of Christmas Beers

Every year in December, Norwegian newspapers report quality tests of Christmas beers. This is a field day for beer enthusiasts. For statistics instructors, it is also an opportunity to collect data for teaching. We may use such data to examine the associational research question of, say, how the bottle price of beers is associated with their alcohol level, if at all. Yet, this is all for later; right now, we are interested in describing the central tendency of the variables' price, alcohol level, and taste, independent of each other. The data x-mas_beer comprise 75 beers (33 cl.). The price per bottle is in Euros, the alcohol level is in percent, and the taste quality has values ranging from zero (terrible) to ten (perfect). If this sounds like 'too much information,' think of a table with three columns (one for each variable) and 75 rows (one for each bottle of beer.) The complete documentation for all the variables in the Christmas beer data appears in appendix A of this chapter.

The mean of the bottle price variable appears in Stata-output 2.5. Since the point-and-click routine in SPSS (Analyze → Descriptive Statistics → Descriptives) of course yields the same results, I see no reason to report this. Such a repeat-everything-in-SPSS approach with no new information of interest will inevitably become very tiresome. The mean or average Christmas beer costs about 6.30 Euros per bottle. All 75 beers in the data lie in the price range from 3.8 to 9.9 Euros.

```
. sum price

    Variable |        Obs        Mean    Std. Dev.        Min        Max
-------------+--------------------------------------------------------
       price |         75        6.28    1.351676        3.8        9.9
```

Stata-output 2.5 Descriptive statistics for the variable `price` in the Christmas beer data.

The frequency distribution for the price variable appears in Stata-output 2.6 and looks similar in SPSS. The median price is 6.1 Euros; note the gray shading for the 52nd cumulative percentage. The mode – that is, the most frequent price among the 75 beers – is 5.5 Euros. Eight beers cost this much (in bold).

Table 2.3 sums up the descriptive statistics for the price variable, the alcohol percent variable, and the taste quality variable. The average beer has an alcohol level of 8.2 percent, a median alcohol level of 8.0 percent, and a mode alcohol level of 9.0 percent. Finally, the average beer scores 5.4 points on the taste quality scale from zero to ten, with a median and a mode of 6.0 points. We need some form of yardstick if we are to make any substantive interpretations of these results, such as similar results for an earlier year. (I promise to stop reiterating this point now!) We will return to these data on several occasions later in the book.

```
. tab price

       price |      Freq.     Percent        Cum.
-------------+-----------------------------------
         3.8 |          1        1.33        1.33
           4 |          2        2.67        4.00
         4.2 |          1        1.33        5.33
         4.5 |          1        1.33        6.67
         4.6 |          3        4.00       10.67
         4.7 |          3        4.00       14.67
         4.8 |          1        1.33       16.00
           5 |          3        4.00       20.00
         5.1 |          1        1.33       21.33
         5.2 |          1        1.33       22.67
         5.3 |          1        1.33       24.00
         5.5 |          8       10.67       34.67
         5.7 |          2        2.67       37.33
         5.9 |          5        6.67       44.00
           6 |          4        5.33       49.33
         6.1 |          2        2.67       52.00
         6.3 |          1        1.33       53.33
         6.5 |          7        9.33       62.67
         6.6 |          3        4.00       66.67
         6.8 |          1        1.33       68.00
```

Stata-output 2.6 Frequency (distribution) table for the variable price in the Christmas beer data.

(*Continued*)

```
       6.9 |          2        2.67        70.67
         7 |          6        8.00        78.67
       7.3 |          1        1.33        80.00
       7.4 |          2        2.67        82.67
       7.5 |          3        4.00        86.67
       7.9 |          1        1.33        88.00
         8 |          2        2.67        90.67
       8.2 |          1        1.33        92.00
       8.7 |          3        4.00        96.00
       9.3 |          1        1.33        97.33
       9.9 |          2        2.67       100.00
-----------+-----------------------------------------
     Total |         75      100.00
```

Stata-output 2.6 (Continued)

Table 2.3 Descriptive statistics for price, alcohol level, and taste quality in the Christmas beer data.

Variable	N =	Mean	Median	Mode
Price per bottle (in Euros)	75	6.3	6.1	5.5
Alcohol level (percent)	75	8.2	8.0	9.0
Taste quality (0–10)	75	5.4	6.0	6.0

2.3 Variables' Measurement Levels: Continuous or Categorical Variables

Three of the variables thus far – infection rate, beer price, and beer alcohol level – have numbers as values. These are continuous or numerical variables – or on the continuous/numerical measurement level or scale.[6] (Some prefer the term ratio variables, but I will not use this term.) It thus makes sense to talk about more or less of the variable in question. Furthermore, phrases like 'twice as many,' 'twice the amount of,' and 'half as much/many' are meaningful. That is, a beer priced at eight Euros costs twice as much as a beer priced at four Euros and so on.

Many variables in the behavioral sciences are not on this measurement level, however. We have for example the variable production location in our Christmas beer data: Such a beer is Norwegian-made *or* made outside of Norway. Furthermore, we have the variable gender in our upcoming student data: A student is either male or female. Both of these variables have a categorical nature, and the categorical variable is the term I will use in this book. Strictly correct, the production location variable and the gender variable are on the nominal measurement level.[7] You may think of nominal variables as a subset of categorical variables if you feel the need to.

The way to classify a variable as categorical in this book is to recognize that it has no ranking among the categories it may take on. (A categorical variable takes on categories, not values.) That is, male category is not more or less than female category; a Norwegian origin is not more or less than a foreign make. Simply put, categorical variables have an either-or logic. The mean and median do not make sense as measures of central tendency for such variables, but the mode very much does. Stata-output 2.7 presents the frequency distribution for the production location variable in the Christmas beer data.

```
. tab country

         country |      Freq.      Percent        Cum.
-----------------+-----------------------------------
          Norway |         47        62.67       62.67
  Outside Norway |         28        37.33      100.00
-----------------+-----------------------------------
           Total |         75       100.00
```

Stata-output 2.7 Frequency table for the variable production location in the Christmas beer data.

There are two categories: made in Norway or made outside Norway. The most typical origin is a Norwegian-made beer, as might be expected given the context: 47 out of the 75 beers, or 63 percent of them, are of Norwegian make. This is the mode. SPSS-output 2.4, where I now only show the basic results, tells the same story, of course.

Table 2.4 displays the information for the production location variable in a more camera-ready publication version.

country

		Frequency	Percent	Valid Percent	Cumulative Percent
Valid	Norway	47	62,7	62,7	62,7
	Outside Norway	28	37,3	37,3	100,0
	Total	75	100,0	100,0	

SPSS-output 2.4 Frequency table for the production location variable in the Christmas beer data.

Table 2.4 illustrates an essential point in passing: The only thing we – or more precisely, the statistics program – may do for a categorical variable is to *count* how many units there are in each category. Categorical variables having only two categories (or outcomes) carry a special name in statistics: *dummy* variables or *dummies*.[8] Yet categorical variables may take on more than two categories, as the next subsection shows.

Table 2.4 Frequency table for the production location variable in the Christmas beer data. $N = 75$.

Variable: Production location	Frequency	Percent
Norway	47	63
Outside of Norway	28	37

2.3.1 The Income and Other Characteristics of Norwegian Soccer Players

Sports superstars make lots of money. The soccer players in the Norwegian version of the Premier League do not belong to this superstars league, but they still command larger earnings than most Norwegians. Previous research has looked at how different factors explain variations in top athletes' earnings, and this is the context for the analyses in Chapters 3 and 4. Now we focus on the typical income and the typicalness of some other attributes of the 240 players in the top-tier Norwegian soccer league. The data are called soccer. The complete documentation for all the variables in the soccer player data appears in Appendix B of this chapter. Summary descriptive statistics for three of the variables in the data appear in Table 2.5. (Nothing new happens regarding what we already have done in Stata or SPSS to produce the results in Table 2.5.)

The soccer players earn about 86,000 Euros on average, whereas their median income is 68,000 Euros. This difference in mean and median by almost 20,000 Euros is huge. Why is this? The answer appears in the so-called histogram in Figure 2.1.

The height of the tallest bars on the left side of the figure shows that most players have an income below 100,000 Euros. (We already know that 50 percent of the players earn less than the median of 68,000 Euros.) On the right side of the figure, that is, for the very low bars, we find the players with very large incomes. These few players' incomes pull the mean income upwards and to the right of the median income – to 86,000 Euros. An old joke starring Bill Gates, at one point the world's wealthiest man, sheds light on this phenomenon. Attuned to the present it goes something like this:

> Jeff Bezos, the now previous CEO of Amazon, walks into a bar. The crowd immediately roars. They all suddenly became billionaires – *on average*.

The joke is not very funny, but has a point. Although the mean wealth among the crowd increased in perhaps billions, the median wealth did not increase by one cent. That is, the mean is more sensitive to extreme values than the median. Against this backdrop, we could claim that the median is a better measure of central tendency than the mean for our soccer income variable. This often happens for distributions of the kind portrayed in Figure 2.1, which we call a skewed distribution. For non-skewed or symmetrical distributions, in contrast, we tend to prefer the mean to the median.[9]

The mode for yearly income is 76,142 Euros. Yet, this mode is uninteresting as a measure of central tendency because it applies to only two out of the 240 players. This is often the case for continuous variables with a large and fine-grained number of values, but not always (as we soon will see).

Table 2.5 Descriptive statistics for yearly income in Euros in 2015, number of club matches played during career, and number of matches played for the national team during career in the soccer player data.

Variable	N =	Mean	Median	Mode
Yearly income in Euros in 2015	240	86,367	67,708	76,142[a]
Club matches in career	240	85	63	29[b]
Nat. team matches in career	240	3.6	0	0[c]

[a] Refers to two players only.
[b] Refers to seven players only.
[c] Refers to 160, or 67 percent, of the players.

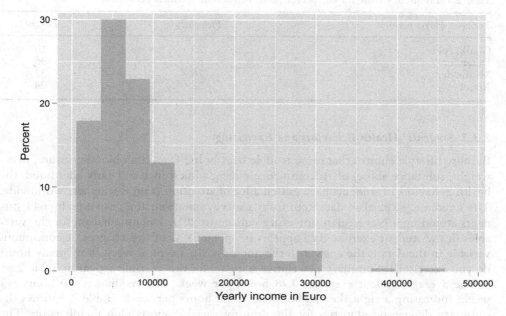

Figure 2.1 Histogram of soccer players' yearly income in Euros in 2015.

The comments for income also apply to club matches during a career: a right-skewed distribution and a mode that does not reveal anything of interest. However, we note an even more pervasive right skew for the variable national team matches during their career: a median of zero and a mean of 3.6. The reason is that a large proportion of the players have *not* played any national team matches at all. A frequency distribution (not shown) shows that this proportion pertains to 160, or 67 percent, of the players. That is, the mode of zero matches played for a national team is in this case an informative descriptive summary measure.

For continuous variables with a lot of skewness, we might consider creating a new variable. For the variable number of matches played for a national team, this could imply making a new variable with only two categories: not having played for a national team at all *or* having played one or more matches for such a team. Table 2.6 presents the summary descriptive statistics for this national team dummy. The mode is having played zero matches for a national team. The actual proportion is 67 percent, as mentioned above.

A categorical variable often has more than two or three categories. A categorical variable with four categories appears in the soccer data. Table 2.7 shows this variable's frequency distribution. We note that the mode is a defense player, with a proportion of 35 percent. The second most-frequent player position is that of a midfielder: 32 percent.[10] We will get back to these soccer data later in the book.

Table 2.6 Frequency table for the national team dummy variable in the soccer data. $N = 240$.

Variable: National team	Frequency	Percent
National team matches = 0	160	67
National team matches > 0	80	33

Table 2.7 Frequency table for the player position variable in the soccer data. $N = 240$.

Variable: Player position	Frequency	Percent
Goalkeeper	23	10
Defender	84	35
Midfielder	76	32
Attacker	57	24

2.3.2 Students' Health Behaviors and Exercising

Because lifestyle choices that cause trouble later in life (e.g., high blood pressure, overweight, substance abuse) often commence during adolescence and early adulthood, the health behaviors of students have gotten a lot of attention from researchers worldwide. The `student_exercise` data stem from a survey questionnaire answered by 644 students attending a Norwegian university college in 2017. Documentation for the variables in the student exercise data appears in appendix C of this chapter. A continuous variable in the data is the answer to the question, 'In a typical week, how many hours do you exercise?'[11] Summary descriptive statistics for this variable appear in Table 2.8.

The average student exercises 4.78 hours per week. The median is 4.0 hours per week, indicating a right skew. The mode is 3.0 hours per week. Table 2.9 shows the summary descriptive statistics for the dummy variable sports club membership. The mode is *not* being a member of a sports club; this is the most frequent category among the students, with a 78 percent proportion.

Table 2.10 shows a frequency distribution table for the variable exercise preference, that is, the answer to a question of what kind of exercising one prefers. Most students prefer to do strength training and cardio training equally. This proportion is 43 percent and is, thus, the mode. We return to these data in the next section.

Table 2.8 Descriptive statistics for hours of weekly exercise in the student exercise data.

Variable	$N =$	Mean	Median	Mode
Hours of weekly exercise	644	4.78	4.0	3.0

Table 2.9 Frequency table for sports club membership in the student exercise data. $N = 644$.

Variable: Member of sports club	Frequency	Percent
No	504	78
Yes	140	22

Table 2.10 Frequency table for exercise preference in the student exercise data. $N = 644$.

Variable: Exercise preference	Frequency	Percent
Strength training	222	34
Cardio training	147	23
Both forms equally much	275	43

2.4 Ordinal Variables: A Third and Special–Case Measurement Level

We find the ordinal variables between the continuous and the categorical variables.[12] Ranking is fine for ordinal variables just as for continuous variables. More or less of this, or higher or lower of that, thus makes sense. Yet exactly how much more or less, or how much higher or lower, is not easy to answer. In accordance with categorical variables, however, ordinal variables take on only a limited number of categories. The examples below clarify.

The Christmas beer data include a taste quality variable; cf. Table 2.3. This variable has 11 categories ranging from zero (terrible taste) to ten (perfect taste). This is ranking. It makes sense to assess a beer scoring an eight as better tasting than a beer scoring a four. Yet it does not make sense to judge the former as twice as tasty as the latter, although eight is twice as much as four. Why? The answer lies in the arbitrariness of the quality scale's end points. That is, the scale could have gone from one to five or from zero to seven. Indeed, the chief taste quality scale for alcoholic beverages, the so-called Parker scale, goes from 60 points (terrible) to 100 points (perfect). The vital point is that we cannot use the distances between scores as measures of how much better or worse beer A tastes compared to beer B or C. This is why, strictly speaking, the taste quality variable is ordinal and not continuous.

I analyzed the quality variable as a continuous variable in Table 2.3. The reason was that it provided a mean (6.4) and a median (6.0). More precisely, I *treated* the variable as continuous to use the mean and median as summary measures of central tendency. Yet there is a more subtle reason for doing this. The mean and median are much more efficient summaries of the typicalness of the taste quality variable than a frequency distribution. The frequency distribution appears in Stata–output 2.8 for comparison. We note that the quality score of three accounts for 23 percent of the beers in the data. Similarly, the score of five accounts for 44 percent, which is still not enough to be the median quality. The median of six, in bold, accounts for 63 percent of the beers in terms of taste quality.

```
. tab quality

   quality |      Freq.      Percent        Cum.
-----------+-----------------------------------
         0 |          8        10.67        10.67
         1 |          1         1.33        12.00
         2 |          5         6.67        18.67
         3 |          3         4.00        22.67
         4 |          9        12.00        34.67
         5 |          7         9.33        44.00
         6 |         14        18.67        62.67
         7 |          8        10.67        73.33
         8 |          7         9.33        82.67
         9 |         11        14.67        97.33
        10 |          2         2.67       100.00
-----------+-----------------------------------
     Total |         75       100.00
```

Stata-output 2.8 Frequency table for the taste quality variable in the Christmas beer data.

Table 2.11 Descriptive statistics for youth sports involvement in the student exercise data.

Variable:	N =	Mean	Median	Mode
Youth sport involvement	644	6.56	7	10

A question in the student exercise data reads, 'In your youth before you started studying, to what extent were you involved in sports requiring lots of physical exercise (to a very small extent = 1; to a very great extent = 10)?' This variable is ordinal: A student answering six was more active than a student answering three, but she was probably not exactly twice as active. Summary descriptive statistics for this variable, which we treat as continuous because of the large number of categories, appears in Table 2.11. Most students answer towards the active end of the scale, and the mode score is ten. A not-shown frequency distribution shows that 127 of the students (20 percent) answered ten.

The two ordinal variables above had many categories: eleven and ten. It is common practice, as I have shown, to treat such variables as continuous in statistical analysis.[13] Yet many ordinal variables have only a few ranked categories, such as five or lower. It is time to look at these.

2.4.1 Ordinal Variables with Few Categories

The questionnaire for the student exercise data includes the question, 'How is your physical health in general?' Table 2.12 provides summary descriptive statistics for the answers to this question.

The health variable is ordinal. The reason is that good health (the mode) is better than ok health, and that very good health is better than both good health and ok health.[14] Ranking thus makes sense. Yet because the variable has only three ranked categories, it makes no sense talking about means and medians. There are typically two main ways of dealing with ordinal variables having few categories in statistical analysis:

(1) Treat the variable as categorical in our sense of the term, that is, to disregard the ordinal nature of the categories or to regroup the variable into two categories, that is, to make it a dummy.
(2) Keep the variable ordinal, and do 'ordinal statistical analysis.'[15]

I mainly adopt category (1) in this book in the spirit of keeping matters as simple as possible. That concludes the basics on variables' measurement levels,[16] and we are ready for a new topic: visual presentations of descriptive statistics.

Table 2.12 Frequency table for physical health assessment in the student exercise data. N = 644.

Variable: Physical health	Frequency	Percent
Ok	217	34
Good	315	49
Very good	112	17

2.5 Visual Presentation of Descriptive Statistics: Graphs

I have used tables to convey statistical information thus far. But since many people think better in visual terms rather than numerically, it is time to do something about that. This section takes on presenting results of descriptive statistical analysis visually. We start with continuous variables and proceed to categorical variables.[17] There are no specific graphs for ordinal variables. In keeping with the comments at the tail end of Section 2.4, I treat ordinal variables as continuous or categorical depending on the situation.

2.5.1 Graphs for Continuous Variables: Boxplots

A natural start for a graph of a continuous variable is the boxplot or box and whiskers plot. Figure 2.2 illustrates for the infection rate variable in the COVID-19 data. The shaded rectangle in the figure – *the box* – contains the infection rates for the middle 50 percent of the districts. The horizontal line inside the box is the median, which we already know is 298. The whiskers are the two remaining horizontal lines; one below and one above the box. The vertical distance from top to bottom whisker accounts for all the infection rates in the data. Any district above or below the whiskers are called *outliers*, of which there are none in the present case.

A similar box and whiskers plot appears in Figure 2.3 for the price variable in the Christmas beer data. We know from Table 2.3 that the median price is 6.1 Euros. Figure 2.3 shows that we find 50 percent of the beers in the price range between about five and a half and seven Euros, that is, within the outer limits of the box. Practically speaking, we find all beers in the price range from about four to about nine Euros, that is, the distance between the two whiskers. Finally, we have three very costly and thus

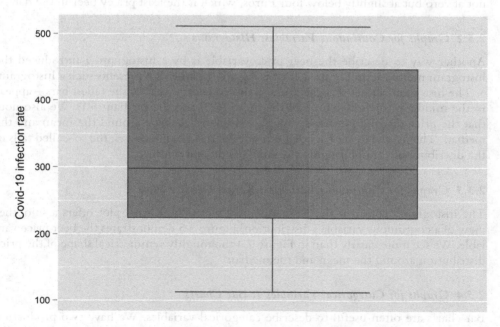

Figure 2.2 Box and whiskers plot (boxplot) for the variable infection rate.

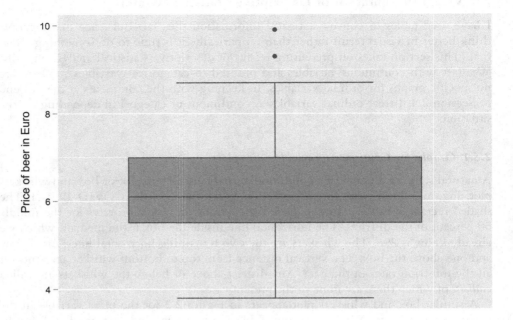

Figure 2.3 Box and whiskers plot (boxplot) for the variable beer price.

outlying beers. Two of these, occupying the same dot in the figure, cost 9.9 Euros, and one costs 9.3 Euros. Note that the axis for price is truncated. That is, the *y*-axis starts not at zero but at slightly below four Euros, which is the least pricey beer in the data.

2.5.2 Graphs for Continuous Variables: Histograms

Another way to describe the beer price variable is by a histogram. I introduced the histogram in passing in Figure 2.1. (It is a graph!) Figure 2.4 presents such a histogram.

The histogram shows that the average-priced beers – that is, the tallest bars – appear in the middle of the graph around the mean (6.3) and the median (6.1). We also note that the price distribution has a roughly symmetrical shape around the mean and the median. That is, the lower bars on both sides of the mean – that is, the so-called tails of the distribution – are of roughly the same height and width.

2.5.3 Graphs for Continuous Variables: Kernel Density Plots

The histogram's cousin is the Kernel density plot. The Kernel plot offers a smoother view of a continuous variable's distribution. Figure 2.5 demonstrates the beer price variable. We see more clearly than in Figure 2.4 the roughly symmetrical shape of the price distribution around the mean and the median.

2.5.4 Graphs for Categorical Variables 1: Bar Charts

Bar charts are often useful to describe categorical variables. We have two production locations for the beers in the Christmas beer data; cf. Table 2.4. Figure 2.6 displays the

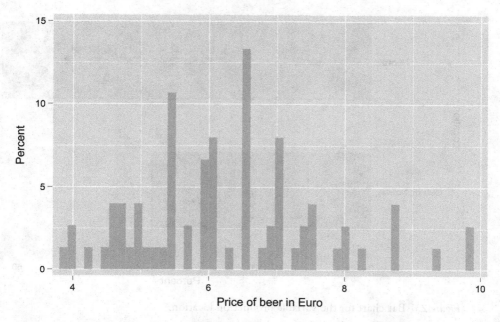

Figure 2.4 Histogram for the variable beer price.

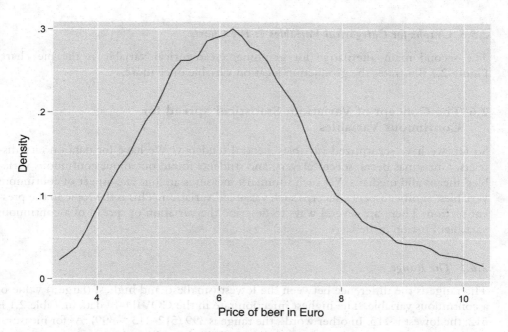

Figure 2.5 Kernel density plot for the variable beer price.

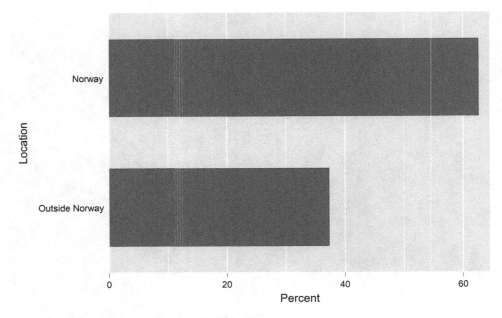

Figure 2.6 Bar chart for the variable production location.

same information as did Table 2.4 but in a visual manner. Most beers (63 percent) are manufactured in Norway. Alternatively, we could place the bars vertically rather than horizontally – this is a matter of taste or convenience.

2.5.5 Graphs for Categorical Variables 2: Pie Charts

The second main alternative for graphing a categorical variable is the pie chart. Figure 2.7 illustrates the production location variable once more.

2.6 The Concept of Variation: Statistical Spread for Continuous Variables

So far, we have scrutinized variables' central tendency. We have for data on city districts, Christmas beers, soccer players, and students found out about continuous variables' means and medians. Yet such summary measures are but one aspect of continuous variables; another aspect concerns such variables' variation. This is the topic in the present section. There are several ways to describe the variation or spread of a continuous variable. First up is the *range*.

2.6.1 The Range

The range is the difference between the lowest (smallest) and highest (largest) value of a continuous variable. The highest infection rate in the COVID-19 data in Table 2.1 is 512; the lowest is 113. In other words, the range is 399 (512–113 = 399). As for measures of central tendency, a range of 399 is not small nor large in itself; we must compare it

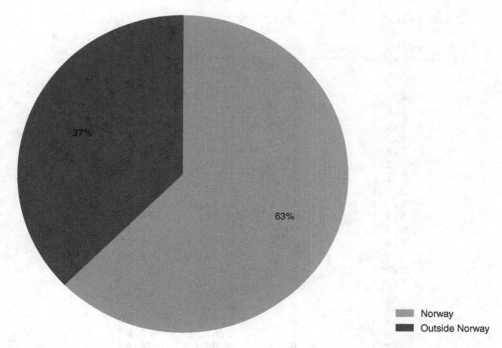

Figure 2.7 Pie chart for the variable production location.

with something. If a group of districts in another city has a range of 200 among their infection rates, we might argue there is more variation in our group of districts. In Stata-output 2.9, which repeats Stata-output 2.6, we see that the least expensive beer in our beer data costs 3.8 Euros, whereas the two most expensive beers cost 9.9 Euros. The range is thus 6.1 Euros.

2.6.2 The Interquartile Range (IQR)

The range has one drawback as a measure of variation: Extreme values influence it. We use the interquartile range (IQR) to account for this. Remember the boxplot in Figure 2.3 containing 50 percent of beers in the middle of the price distribution. These beers similarly appear between the row in bold (**5.5**) and the row in bold italics (**7.0**) in Stata-output 2.9. The IQR is the price range of the beers inside this 50-percent box: 1.5 Euros (7.0–5.5 = 1.5).

Some of the students in our exercise data exercise for zero hours per week (result not shown). At the other end of the spectrum, some students exercises for 16 hours per week. The range is thus 16. In contrast, the IQR pertaining to the 50 percent of students in the middle of the exercise distribution is only three hours. Yet there is no point showing this because it is just a long table akin to Stata-output 2.9. This concludes our business with the range as a measure of a variable's spread or variation. Next up is the standard deviation (SD).

```
. tab price

      price |      Freq.      Percent         Cum.
------------+-----------------------------------------
        3.8 |          1         1.33         1.33
          4 |          2         2.67         4.00
        4.2 |          1         1.33         5.33
        4.5 |          1         1.33         6.67
        4.6 |          3         4.00        10.67
        4.7 |          3         4.00        14.67
        4.8 |          1         1.33        16.00
          5 |          3         4.00        20.00
        5.1 |          1         1.33        21.33
        5.2 |          1         1.33        22.67
        5.3 |          1         1.33        24.00
        5.5 |          8        10.67        34.67
        5.7 |          2         2.67        37.33
        5.9 |          5         6.67        44.00
          6 |          4         5.33        49.33
        6.1 |          2         2.67        52.00
        6.3 |          1         1.33        53.33
        6.5 |          7         9.33        62.67
        6.6 |          3         4.00        66.67
        6.8 |          1         1.33        68.00
        6.9 |          2         2.67        70.67
          7 |          6         8.00        78.67
        7.3 |          1         1.33        80.00
        7.4 |          2         2.67        82.67
        7.5 |          3         4.00        86.67
        7.9 |          1         1.33        88.00
          8 |          2         2.67        90.67
        8.2 |          1         1.33        92.00
        8.7 |          3         4.00        96.00
        9.3 |          1         1.33        97.33
        9.9 |          2         2.67       100.00
------------+-----------------------------------------
      Total |         75       100.00
```

Stata-output 2.9 Frequency (distribution) table for the variable price in the Christmas beer data.

2.6.3 The Standard Deviation

Some of the districts in the COVID-19 data have infection rates close to the overall mean of 317, such as 311 (district C) or 351 (district G). Others have rates further away from the mean – say 224 or 448. That is, the individual districts differ regarding the deviation between their own infection rate and the (overall) mean infection rate among all the 15 districts. Consider district C in Table 2.1 (311). The deviation or distance from the overall mean for this district is eight (317–311 = 8). For district G, with a rate of 351, the similar deviation is 34. Fast-forward to the key idea: Because every district has a deviation from the overall mean in the data, we may compute a measure resembling the average deviation from the overall mean. Please welcome the *standard deviation* or SD.

Technically, the standard deviation is the square root of a measure called the *variance*. We obtain the variance or s^2 for the infection rate variable, now labeled as y, by the formula

$$s^2 = \frac{\sum (y_i - \bar{y})^2}{n-1},$$

where $\sum$ is the summation sign, y_i is the individual district's infection rate, $\bar{y}$ is the mean infection rate in the data, and n is the number of districts in the data. The variance or s^2 is 14,592.22 in this case. The square root of the variance, or s, is the SD: 120.80. The SD measures the size of the variation or spread around the mean of a continuous variable: The larger the SD, the more variation around the mean there is. The practical problem once more is that the size of a SD is neither small nor large in itself; it needs context for assessment, such as the SD for the infection rates in another group of city districts. If the latter group has a SD of, say, 250, we could claim there is less variation among the infection rates in Oslo. Having explained how SD works, we are finally ready to present the relevant summary descriptive statistics for continuous variables.

2.6.4 Presenting Descriptive Statistics for Continuous Variables in a Thesis or a Research Paper

Table 2.13 is a typical way of showing summary descriptive statistics for a continuous variable using the familiar infection rate variable as an example. (We do not normally present the IQR.) Table 2.14 shows the similar setup for the two continuous variables in the Christmas beer data.

Consider the SDs in Table 2.14. It is tempting to claim that the variation in prices is smaller than the variation in alcohol levels: 1.35 versus 1.57. This is not necessarily the case, however, because we cannot interpret the SD in this way. Since the two variables refer to different measurement scales (i.e., Euros and percent), comparing the two SD amounts is akin to comparing apples with oranges. We use the coefficient of variation (CV) to overcome this problem.

Table 2.13 Descriptive statistics for the infection rate variable in the COVID-19 data.

Variable:	N =	Mean	SD	Median	Min.	Max.
Infection rate	15	317	121	298	113	512

Table 2.14 Descriptive statistics for price and alcohol level in the Christmas beer data.

Variable:	N =	Mean	SD	Median	Min.	Max.
Price per bottle (Euros)	75	6.3	1.35	6.1	3.8	9.9
Alcohol level (percent)	75	8.2	1.57	8	5	12

2.6.5 Coefficient of Variation

Finding the CV is straightforward. We divide the SD by the mean and multiply the result with 100. For the price and alcohol variable, respectively, we thus obtain 21.4 and 19.1. Since these two CVs are comparable, the correct interpretation is that the price variable in fact has more variation than the alcohol level variable.

2.6.6 A Few Remarks on the Shape of Continuous Variables' Distribution

We often learn a lot from knowing the central tendency and the variation of a continuous variable. Yet these two pieces of information do not necessarily say much about the *shape* of such a variable's distribution. The next few paragraphs introduces this topic in a tentative fashion.

The alcohol level variable in Table 2.14 has a range of seven (12–5 = 7) and a mean of 8.2 percent. The histogram in Figure 2.8 shows this visually. Superimposed on the figure is the normal or Gaussian distribution with its characteristic bell-shaped and perfectly symmetric form.[18] We know *a priori* – that is, before doing any analysis at all – that the mean, median, and the mode are similar for a variable having a normal distribution. We may deduce the following from this:

- The alcohol level variable is not normally distributed, but it comes rather close as a practical matter
- The alcohol level variable has a symmetric form with little skewness to either side
- The peak (i.e., top) of the alcohol level variable's distribution is less sharp than for the normal distribution

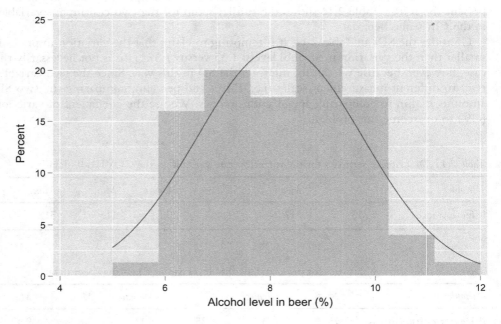

Figure 2.8 Histogram for the variable alcohol level, with the normal distribution superimposed.

Skewness and *kurtosis* measure the characteristics of a continuous variable's distribution against the normal distribution. A perfect normal distribution has zero skewness and a kurtosis of three. The alcohol variable has a skewness of 0.092, thus making it symmetrical for all practical purposes.[19] The kurtosis is 2.18. This suggests that the peak of the distribution is less sharp than for the normal distribution, and that the tails on both sides are thinner and shorter than for the normal distribution.[20]

Why bring this up? The normal distribution plays a fundamental role in inferential statistics, that is the topic of Chapter 5 and Section 5.3 in particular. This was just a small teaser in terms of preparation.

2.7 Foreshadowing Associational Research Questions: Descriptive Statistics for Subgroups

To set the stage for Chapter 3, we close with a section on how two variables might be associated in some way. We look at the price variable and the production location variable in the Christmas beer data to illustrate. Stata-output 2.10 lists a number of summary statistics measures for the price variable broken down on the two production locations: in Norway or outside of Norway. Yet since results like these typically are intermediate in the research process, I see no reason to present a camera-ready publication table for the results. (The results are of course similar in SPSS.)

```
. tabstat price, by(country) stats(count min max range mean sd cv)

Summary for variables: price
    by categories of: country

    country |      N      min      max    range     mean       sd        cv
------------+--------------------------------------------------------------
     Norway |     47        4      9.9      5.9  6.389362  1.529313  .239353
Outside Norway |  28      3.8      7.5      3.7  6.096429  .9833737  .1613032
------------+--------------------------------------------------------------
      Total |     75      3.8      9.9      6.1     6.28  1.351676  .215235
---------------------------------------------------------------------------
```

Stata-output 2.10 Summary descriptive statistics for the variable price in the Christmas beer data, by production location (country).

The bottom row (`Total`) provides the same information as did Table 2.14. Other than this, we note that the mean prices are close for the two production locations: 6.4 Euros for Norwegian-made beers and 6.1 Euros for beers made outside of Norway. In contrast, there seems to be more price variation among the Norwegian beers because the range, the `sd`, and the `cv` are all larger for this group of beers. Figure 2.9 paints the same picture as Stata-output 2.10.[21] It appears, in other words, as if the price variable and the production location variable are associated somehow. I have much more to say on such variable associations in Chapter 3.

2.8 Chapter Summary, Key Learning Points, and Further Reading

This chapter has been about doing statistical analysis and presenting summary descriptive statistics for one variable at a time. Descriptive statistical analysis primarily concerns

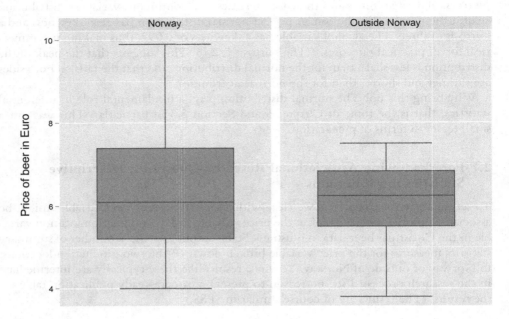

Figure 2.9 Box and whiskers plot (boxplot) for the variable beer price, by production location.

finding out about what is typical for variable *x* and variable *y,* and so on. Yet descriptive statistical analysis is also about finding the variation or spread of a continuous *x* and *y* and so on. Below follows some key learning points:

- Three statistical summary measures express the central tendency or typicalness of a variable: the mean (or average), the median, and the mode.
- There are several ways of categorizing a variable according to its measurement levels. I use two main levels in this book: the continuous measurement level (i.e., continuous variables) and the categorical measurement level (i.e., categorical variables).
- Ordinal variables (i.e., variables at the ordinal measurement level) are a special case of variables located between the continuous variables and the categorical variables.
- Means and medians make sense as summary measures of continuous variables and ordinal variables having many categories, say typically six or more. The mode, in principle, applies to all measurement levels as a summary measure.
- The median is often a better descriptive summary measure of a variable than the mean for skewed distributions.
- The results of descriptive statistics may be presented in tables or in graphs. Or both.
- Statistical variation or spread primarily concern continuous variables. The most important measure in this regard is the SD.

Freedman et al. (2007) and Agresti (2018) are general books on statistics/statistical analysis teaching you all there is to know. Two introductions on doing statistical analysis in Stata are Bittmann (2019) and Daniels and Minot (2020), whereas Field (2018) is the key source in this regard for SPSS. Wheelan (2013) and Spiegelhalter (2019) are two entertaining introductions to statistics and statistical analysis. Healy (2019) is a good starting-point for data visualizations (i.e., graphs).

2.9 Executing Statistical Commands: Do-Files in Stata and Syntax-Files in SPSS

After a session of interactive statistical analysis, Stata or SPSS will ask if you want to save the changes you made (if any) to the data. If you consider answering yes to this question, I recommend that you save the data under a new name. In this way, you keep your original data intact if anything has gone wrong during the session. (It often does!) Working interactively in this way is not the favored approach among instructors, however. No, they – and I – urge you to work commando-style and save only the commands you made during the session. This way, the data are left unchanged. Please welcome do-files and syntax-files, two ways of working commando-style with statistical analysis.

A do-file is a text document containing the commands you tell Stata to execute during a session. Similarly, a syntax-file is a text document containing the commands you tell SPSS to execute. The reason instructors advocate using do-files/syntax-files is that they ensure *reproducibility* and *translucency* in research. Translation: We forget what we did yesterday or three hours ago in our interactive sessions!

2.9.1 Stata Commands in Do-Files

A Stata do-file is a text file containing the commands, some of which we have seen so far, such as `sum infect_rate` or `tab quality`. Instead of typing such commands in the Command-window, however, we type them in a text document and execute them – or 'do them' – from within this document; hence the name do-file. For example, to obtain the mean and the frequency distribution for the infection rate variable in the COVID-19 data (Stata-outputs 2.3 and 2.4), I typed the following in a do-file:

```
version 16.1
capture log close
set more off
use "C:\Users\700245\OneDrive - Høgskolen i Innlandet\Dokumenter\
Christer\Forskning\Statistical_analysis_research_persp\res_prop_
price_Covid-19.dta"
sum infect_rate
tab infect_rate
```

The text above appears in do-file format in Stata-output 2.11. I highlighted lines 1 to 8 and clicked on the execute-button in the upper-right corner (pointed at by the arrow) to execute or 'do' the commands.[22]

Line 5, stretching out over three lines, tells Stata where I have stored the data on my computer. The location for the data on yours will of course be different. You may copy-paste the analogous text on your computer from the Results-window the first time you open the data in interactive mode. Lines 7 and 8, the two statistical commands, produce the results we saw in Stata-outputs 2.3 and 2.4.

Below the `tab`-command to get the frequency distribution for `infect_rate`, one may continue the do-file with more commands or with a new `use`-line to get, say, the Christmas beer data:

Stata-output 2.11 An example of a short Stata do-file.

```
use "C:\Users\700245\OneDrive - Høgskolen i Innlandet\Dokumenter\
Christer\Forskning\Statistical_analysis_research_persp\x-mas_beer.
dta", replace
```

After this, one may ask for the mean of the beer price variable with the command:

```
sum price
```

Below you find the Stata-commands to generate the results in this chapter. I present the commands in plain text and not in the do-file format of Stata-output 2.11 to save space. I also skip the long and tedious use-commands (which are probably different on your computer than they are on mine anyway) that always must appear before a command whenever we shift from one data set to another.

Stata-output 2.3 (as mentioned)

```
sum infect_rate
```

Stata-output 2.4 (as mentioned)

```
tab infect_rate
```

Table 2.2
Stata-outputs 2.3 and 2.4 generate the results of this table.

Stata-output 2.5

```
sum price
```

Stata–output 2.6

```
tab price
```

Table 2.3

```
sum price alch_perc quality
tab price
tab alch_perc
tab quality
```

Stata–output 2.7 and Table 2.4

```
tab country
```

Table 2.5

```
sum inc_year match_tot national, detail
tab inc_year
tab match_tot
tab national
```

The extension `detail` after the comma displays the median directly; cf. the 50 percent.

Figure 2.1

```
twoway histogram inc_year, percent
```

Note that the figures in the book containing numerical information look slightly different from Stata's default graphs. Since I prefer a visual layout other than the default graphs, I use a different graph scheme. To download it, go to https://github.com/mdroste/stata-scheme-modern and follow the instructions. I have also done some tinkering to get the graphs camera-ready. Such tinkering is always required.

Table 2.6

```
tab nation_dum
```

Table 2.7

```
tab pos
```

Table 2.8

```
sum hours_exer, detail
tab hours_exer
```

Table 2.9

```
tab sport_club
```

Table 2.10

```
tab exer_most
```

Stata–output 2.8

```
tab quality
```

Table 2.11

```
sum youth_exe, detail
tab youth_exe
```

Table 2.12

```
tab health
```

Figure 2.2

```
graph box infect_rate
```

Figure 2.3

```
graph box price
```

Figure 2.4

```
twoway histogram price, percent bin(50)
```

Figure 2.5

```
kdensity price
```

Figure 2.6

```
catplot country, percent
```

Before applying the `catplot`-command, you must first download it from the Internet. In the Command-window or do-file, type:

```
ssc install catplot
```

Figure 2.7

```
graph pie, over(country) plabel(_all percent)
```

Stata–output 2.9

```
tab price
```

Table 2.13

```
sum infect_rate, detail
```

Table 2.14

```
sum price alch_perc, detail
```

Figure 2.8

```
histogram alch_perc, percent norm
```

The command to obtain the values for skewness and kurtosis is:

```
sum alch_perc, detail
```

Stata output 2.10

```
tabstat price, by(country) stats(count min max range mean sd cv)
```

If you only want the information in the Total row, delete by(country) from the command above.

Figure 2.9

```
graph box price, by(country)
```

2.9.2 SPSS Commands in Syntax-Files

A syntax-file is the SPSS-version of a Stata do-file. Compared to Stata's short commands, however, the syntaxes of SPSS are longer and thus more tedious to type. Thankfully, copy-paste works fine and saves a lot of typing/time. To get the mean and the frequency distribution for the infection rate variable in the COVID-19 data (SPSS-outputs 2.2 and 2.3), I typed the following in a syntax-file:

```
GET
  FILE='C:\Users\700245\OneDrive - Høgskolen i '+
    'Innlandet\Dokumenter\Christer\Forskning\
Statistical_analysis_research_persp\res_prop_price_Cov'+
    'id-19.sav'.
DESCRIPTIVES VARIABLES=infect_rate
  /STATISTICS=MEAN STDDEV MIN MAX.
FREQUENCIES VARIABLES=infect_rate
  /ORDER=ANALYSIS.
```

Yet there is a simpler way! When opening the COVID-19 data (File → Open → Data), do *not* click on Open but on Paste instead. Similarly, after having dragged the variable infect_rate from the box on the left and over to the empty box on the right to get its mean and frequency distribution (two separate commands), do *not* click on OK but on Paste instead. Voilà! You now have the syntax-file in SPSS-output 2.5 without typing anything (yet with a different `Get File` command). Moreover, you may paste all interactive SPSS-commands into a syntax-file in this way. (Stata has a similar capability, but given its very short commands there is not much to gain.) I highlighted lines 2 to 11 and clicked on the execute-button (the triangle, pointed at by the arrow) to execute the commands.

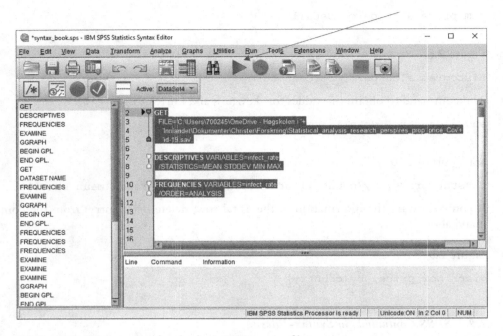

SPSS-output 2.5 An example of a short SPSS syntax-file.

Lines 2–5 tells SPSS where the data are stored on my computer; this location will be different on your computer. Lines 7, 8 and 10, 11, the two statistical commands, produce the results we saw in SPSS-outputs 2.2 and 2.3.

Below the `FREQUENCIES`-command to get the frequency distribution for `infect_rate`, one may continue the syntax-file with more commands or with a new `GET`-line to get, say, the Christmas beer data:

```
GET
  FILE='C:\Users\700245\OneDrive - Høgskolen i '+
    'Innlandet\Dokumenter\Christer\Forskning\Statistical_analysis_
research_persp\x-mas_beer.sav'.
DATASET NAME DataSet2 WINDOW=FRONT.
```

After this, one may ask for the mean of the beer price variable with the command:

```
DESCRIPTIVES VARIABLES=price
  /STATISTICS=MEAN STDDEV MIN MAX.
```

Below you find the SPSS-commands to generate the results in this chapter in SPSS style. I have pasted the commands from the interactive mode as just described. I do not present the commands in the syntax-file format of SPSS-output 2.5 but only as plain text to save space. I also skip the long and tedious `GET`-commands (which are different on your computer than they are on mine anyway) that always must appear before a command whenever you shift from one data set to another.

SPSS-output 2.2 (as mentioned)
```
DESCRIPTIVES VARIABLES=infect_rate
  /STATISTICS=MEAN STDDEV MIN MAX.
```

SPSS-output 2.3 (as mentioned)
```
FREQUENCIES VARIABLES=infect_rate
  /ORDER=ANALYSIS.
```

Table 2.2
SPSS-outputs 2.2 and 2.3 generate the results of this table.

Stata-output 2.5
```
DESCRIPTIVES VARIABLES=price
  /STATISTICS=MEAN STDDEV MIN MAX.
```

Stata-output 2.6
```
FREQUENCIES VARIABLES=price
  /ORDER=ANALYSIS.
```

Table 2.3
```
DESCRIPTIVES VARIABLES=price alch_perc quality
  /STATISTICS=MEAN STDDEV MIN MAX.
FREQUENCIES VARIABLES=price alch_perc quality
  /ORDER=ANALYSIS.
```

SPSS-output 2.4 and Table 2.4
```
FREQUENCIES VARIABLES=country
  /ORDER=ANALYSIS.
```

Table 2.5
```
DESCRIPTIVES VARIABLES=inc_year match_tot national
  /STATISTICS=MEAN STDDEV MIN MAX.
FREQUENCIES VARIABLES=inc_year match_tot national
  /ORDER=ANALYSIS.
```

Figure 2.1
```
FREQUENCIES VARIABLES=inc_year
  /HISTOGRAM
  /ORDER=ANALYSIS.
```

Table 2.6
```
FREQUENCIES VARIABLES=nation_dum
  /ORDER=ANALYSIS.
```

Table 2.7

```
FREQUENCIES VARIABLES=pos
  /ORDER=ANALYSIS.
```

Table 2.8

```
DESCRIPTIVES VARIABLES=hours_exer
  /STATISTICS=MEAN STDDEV MIN MAX.
FREQUENCIES VARIABLES=hours_exer
  /ORDER=ANALYSIS.
```

Table 2.9

```
FREQUENCIES VARIABLES=sport_club
  /ORDER=ANALYSIS.
```

Table 2.10

```
FREQUENCIES VARIABLES=exer_most
  /ORDER=ANALYSIS.
```

Stata-output 2.8

```
FREQUENCIES VARIABLES=quality
  /ORDER=ANALYSIS.
```

Table 2.11

```
DESCRIPTIVES VARIABLES=youth_exe
  /STATISTICS=MEAN STDDEV MIN MAX.
FREQUENCIES VARIABLES=youth_exe
  /ORDER=ANALYSIS.
```

Table 2.12

```
FREQUENCIES VARIABLES=health
  /ORDER=ANALYSIS.
```

Figure 2.2 (basic)

```
EXAMINE VARIABLES=infect_rate
  /PLOT=BOXPLOT
  /STATISTICS=NONE
  /NOTOTAL.
```

Note that all graph commands in SPSS (as in Stata or whatever statistics program you are using) produce very basic graphs that need some tinkering to become camera-ready.

Figure 2.2 (advanced)

```
GGRAPH
  /GRAPHDATASET NAME="graphdataset" VARIABLES=infect_rate MISS-
ING=LISTWISE REPORTMISSING=NO
  /GRAPHSPEC SOURCE=INLINE.
BEGIN GPL
  SOURCE: s=userSource(id("graphdataset"))
  DATA: infect_rate=col(source(s), name("infect_rate"))
  DATA: id=col(source(s), name("$CASENUM"), unit.category())
  COORD: rect(dim(1), transpose())
  GUIDE: axis(dim(1), label("infect_rate"))
  GUIDE: text.title(label("1-D Boxplot of infect_rate"))
  ELEMENT: schema(position(bin.quantile.letter(infect_rate)),
label(id))
END GPL.
```

The GGRAPH-commands originate from the interactive chart-builder function in SPSS, and I have used the paste-procedure in the usual manner. If you type 'chart builder SPSS' on Google search, you will find many videos demonstrating how to use the chart-builder function. See also Field (2018).

Figure 2.3

```
EXAMINE VARIABLES=price
  /PLOT=BOXPLOT
  /STATISTICS=NONE
  /NOTOTAL.
```

Figure 2.4

```
FREQUENCIES VARIABLES=price
  /HISTOGRAM
  /ORDER=ANALYSIS.
```

Figure 2.5

```
GGRAPH
  /GRAPHDATASET NAME="graphdataset" VARIABLES=price MISSING=LISTWISE
REPORTMISSING=NO
  /GRAPHSPEC SOURCE=INLINE.
BEGIN GPL
  SOURCE: s=userSource(id("graphdataset"))
  DATA: price=col(source(s), name("price"))
  GUIDE: axis(dim(1), label("price"))
  GUIDE: axis(dim(2), label("Density"))
  SCALE: linear(dim(2), min(-5))
  ELEMENT: line(position(density.kernel.epanechnikov(price*1)))
END GPL.
```

Figure 2.6 (with vertical bars)

```
FREQUENCIES VARIABLES=country
  /BARCHART FREQ
  /ORDER=ANALYSIS.
```

Figure 2.7

```
FREQUENCIES VARIABLES=country
  /PIECHART FREQ
  /ORDER=ANALYSIS.
```

Stata-output 2.9

```
FREQUENCIES VARIABLES=price
  /ORDER=ANALYSIS.
```

Table 2.13

```
DESCRIPTIVES VARIABLES=infect_rate
  /STATISTICS=MEAN STDDEV MIN MAX.
FREQUENCIES VARIABLES=infect_rate
  /ORDER=ANALYSIS.
```

Table 2.14

```
DESCRIPTIVES VARIABLES=price alch_perc
  /STATISTICS=MEAN STDDEV MIN MAX.
FREQUENCIES VARIABLES=price alch_perc
  /ORDER=ANALYSIS.
```

Figure 2.8

```
FREQUENCIES VARIABLES=alch_perc
  /HISTOGRAM NORMAL
  /ORDER=ANALYSIS.
```

The command to produce the values for skewness and kurtosis is:

```
EXAMINE VARIABLES=alch_perc
  /PLOT BOXPLOT STEMLEAF
  /COMPARE GROUPS
  /STATISTICS DESCRIPTIVES
  /CINTERVAL 95
  /MISSING LISTWISE
  /NOTOTAL.
```

Note that Stata and SPSS use two different formulas to compute the kurtosis. This explains the different results in this regard.[23]

Stata-output 2.10 and Figure 2.9

```
EXAMINE VARIABLES=price BY country
  /PLOT BOXPLOT STEMLEAF
```

```
/COMPARE GROUPS
/STATISTICS DESCRIPTIVES
/CINTERVAL 95
/MISSING LISTWISE
/NOTOTAL.
```

If you do not want the results broken down on country, delete By country in the first line of the command.

2.10 Chapter Exercises with Solutions

The exercises below use the data available for download on the book's website.

Exercises:

Exercise 1 (data: res_prop_price_COVID-19, see Section 2.2 for the data display)
  What is the mean of the variable residential property price per square-meter (price_sq_m) for the 15 city districts? The minimum? The maximum? The range? The median? The SD? The CV?

Exercise 2 (data: soccer, see appendix B of this chapter for data documentation)
  2a What is the mean of the variable number of matches played during the season (match_ses)? The minimum? The maximum? The range? The median? The SD? The CV?
  2b What is the mode for number of goals scored during the season (goals)?
  2c What is the mode for player origin (origin)?

Exercise 3 (data: student_exercise, see appendix C of this chapter for data documentation)
  3a What is the mean of the variable number of times exercising per week (times_exer)? The minimum? The maximum? The range? The median? The SD? The CV?
  3b What is the mode for financial situation (econ)?
  3c How many students are younger than 22 years (age)?

Answers to exercises (in Stata only; see Section 2.9 for equivalent SPSS syntaxes):

Exercise 1 (data: res_prop_price_COVID-19, see Section 2.2 for the data display)
  We could apply the sum and tab commands, but it is faster to use the flexible tabstat-command from Section 2.7 as in:

```
. tabstat price_sq_m, stats(count mean min max range median sd cv)

    variable |        N       mean        min        max      range        p50
-------------+------------------------------------------------------------------
   price_sq_m |       15   67.62673      43.21     90.332     47.122     69.166
-------------+------------------------------------------------------------------

    variable |       sd         cv
-------------+--------------------
   price_sq_m | 15.34453   .2269003
-------------+--------------------
```

The mean is 67.63 in thousands of NOK (or 67,626 230 NOK). The minimum is 43.21, the maximum is 90.33, and the range is thus about 47. The median (p50) is 69.17 or quite similar to the mean. The SD is 15.34 and the CV is 0.23.

Exercise 2 (data: soccer, see appendix B of this chapter for data documentation)

2a   What is the mean of the variable number of matches played during the season (match_ses)? The minimum? The maximum? The range? The median? The SD? The CV?

We could apply the sum and tab commands, but again it is faster to use the tabstat-command from Section 2.7 as in:

```
. tabstat match_ses, stats(count mean min max range median sd cv)

    variable |        N      mean       min       max     range       p50
-------------+--------------------------------------------------------------
   match_ses |      240  20.18333         2        30        28        23
----------------------------------------------------------------------------

    variable |       sd        cv
-------------+--------------------
   match_ses | 8.329494  .4126917
----------------------------------
```

The mean is 20 matches, the minimum is 2, the maximum is 30, and the range is thus 28. The median (p50) is 23 or rather close to the mean. The SD is 8.33 and the CV is 0.41.

2b   What is the mode for number of goals scored during the season (goals)?

Here, we simply use the tab command as in:

```
. tab goals

      goals |      Freq.     Percent        Cum.
------------+-----------------------------------
          0 |         96       40.00       40.00
          1 |         47       19.58       59.58
          2 |         29       12.08       71.67
          3 |         13        5.42       77.08
          4 |         20        8.33       85.42
          5 |          9        3.75       89.17
          6 |          8        3.33       92.50
          7 |          2        0.83       93.33
          8 |          3        1.25       94.58
          9 |          7        2.92       97.50
         10 |          3        1.25       98.75
         11 |          1        0.42       99.17
         13 |          1        0.42       99.58
         15 |          1        0.42      100.00
------------+-----------------------------------
      Total |        240      100.00
```

The mode is zero goals; 40 percent of the soccer players have not scored a goal during the season.

2c   What is the mode for player origin (origin)?

Again, we use the tab command as in:

```
. tab origin
```

origin	Freq.	Percent	Cum.
Norwegian	175	72.92	72.92
foreign	65	27.08	100.00
Total	240	100.00	

The mode is a Norwegian player, with a percentage of 73.

Exercise 3 (data: `student_exercise`, see appendix C of this chapter for data documentation)

    3a What is the mean of the variable number of times exercising per week (`times_exer`)? The minimum? The maximum? The range? The median? The SD? The CV?

We use the `tabstat` command from Section 2.7 as in:

```
. tabstat times_exer, stats(count mean min max range median sd cv)
```

variable	N	mean	min	max	range	p50
times_exer	644	3.038043	0	8	8	3

variable	sd	cv
times_exer	1.652002	.5437716

The mean is about three times per week, the minimum is 0, the maximum is 8, and the range is thus 8. The median (p50) is 3 or very similar to the mean. The SD is 1.65 and the CV is 0.54.

    3b What is the mode for financial situation (`econ`)?

We use the `tab` command as in:

```
. tab econ
```

econ	Freq.	Percent	Cum.
not good	82	12.73	12.73
ok	297	46.12	58.85
good	265	41.15	100.00
Total	644	100.00	

The mode is ok health, with a percentage of 46.

    3c How many students are younger than 22 years (age)?

Again, we use the tab command as in:

```
. tab age

       age |      Freq.      Percent        Cum.
-----------+-----------------------------------
        19 |        50         7.76         7.76
        20 |       114        17.70        25.47
      20.5 |         1         0.16        25.62
        21 |       123        19.10        44.72
        22 |       116        18.01        62.73
        23 |        88        13.66        76.40
        24 |        46         7.14        83.54
        25 |        25         3.88        87.42
        26 |        20         3.11        90.53
        27 |        15         2.33        92.86
        28 |         6         0.93        93.79
        29 |        10         1.55        95.34
        30 |         7         1.09        96.43
        31 |         2         0.31        96.74
        32 |         4         0.62        97.36
        33 |         2         0.31        97.67
        34 |         2         0.31        97.98
        35 |         2         0.31        98.29
        36 |         2         0.31        98.60
        37 |         1         0.16        98.76
        38 |         1         0.16        98.91
        40 |         1         0.16        99.07
        41 |         1         0.16        99.22
        42 |         1         0.16        99.38
        43 |         2         0.31        99.69
        44 |         1         0.16        99.84
        45 |         1         0.16       100.00
-----------+-----------------------------------
     Total |       644       100.00
```

We find that 44.72 percent of the students are younger than 22 years if we look at the cumulative percentage column on the right (Cum.). Note that one student found it necessary to be more specific and rather than answering with an integer, he or she answered 20.5 years!

Notes

1 I write the names of the data sets used in the book in the font Courier New. The same goes for commands and (most) variable names. For example, the COVID-19 infection rate variable bears the name infect_rate.
2 The text appearing in the Results-window in the middle of Stata-output 2.2 shows where the data is stored on my computer. This location will of course be different on your computer.
3 I use the (arithmetic) mean and the average as synonymous and thus interchangeable throughout the book.
4 Note that the command generating the output (tab, short for tabulate) appears on top of Stata-outputs, where you should disregard the dot in front of the command. In this case, a more direct approach to get the median is to use the command: sum infect_rate, detail
5 Note that tables of summary statistics normally include more descriptive measures. We will get back to these in Section 2.6.

6 The taste quality variable is ordinal – or on the ordinal measurement level. This is the topic of Section 2.4. There are several ways of characterizing and presenting variables' measurement levels. My personal touch on the subject follows my aim of keeping matters as simple as possible without sacrificing (too much) precision.

7 These two particular variables are so-called dummy variables, which I will return to shortly.

8 All statistics programs require numbers to do calculations. We thus must provide the categories of a dummy with a pair of numbers. The conventional coding in this regard is 1 and 0. It is customary to code the presence of something or yes as 1 and the non-presence of something or no as 0.

9 Figure 2.1 is a right-skewed distribution; the mean is located on the right side of the median. The long tail to the right in the figure – that is, the very large incomes pertaining to only a few players – characterizes such a distribution. A left-skewed distribution has the mean on the left of the median, with a tail to the left. For distributions with no skewness, that is, symmetrical distributions, the mean and the median are often very close. I will say some more on continuous variables' distributions in Section 2.6.

10 The codings in the data are goalkeeper = 0, defender = 1, midfielder = 2, and attacker = 3; cf. the data documentation. These codings could be anything, however, as the only requirement is four different numbers.

11 One question typically equals one variable when data stem from survey questionnaires.

12 Some will place the ordinal variables among the categorical variables. My reason for not doing so in this book is presentational ease.

13 I often do the same for an ordinal variable having about six or more categories if its frequency distribution is not very skewed.

14 The health variable could have been a scale ranging from zero (very poor heath) to a hundred (excellent health). This would make it continuous in practical terms.

15 The definitive source on ordinal statistical analysis is Agresti (2010).

16 Actually, it does not. There is one more measurement level: the interval level. Yet since interval variables for all practical purposes are similar to continuous variables, I do not address them. I do not think you should do so either.

17 The commands for generating the graphs in this section all appear in Section 2.9.

18 Few real-life variables have perfect normal distributions. Peoples' height and IQ are oft-mentioned examples in this regard.

19 Positive values suggest a right skew, whereas negative values imply a left skew. Values below −1.0 or above 1.0 indicate severe skewness. Values around −0.75 and 0.75 suggest moderate skewness.

20 A kurtosis above 3.0 suggests a sharper peak and that the tails on both sides are thicker and longer than that of the normal distribution.

21 Note, however, that the median price is actually higher for non-Norwegian beers, whereas it is the other way round for the mean price; cf. Stata-output 2.10.

22 Line 1 tells Stata which version of the program one uses (Stata version 17 became available in the spring of 2021); lines 2 and 3 are just smart to enter in every do-file for reasons that we need not get into.

23 See https://stats.idre.ucla.edu/other/mult-pkg/faq/general/faq-whats-with-the-different-formulas-for-kurtosis/ for different formulas.

Appendix A
Christmas Beer Data

Data documentation for the data x-mas_beer; a data set of Christmas beers quality-tested by a Norwegian newspaper in 2019. Variable names are in **bold** typeface. N = 75.

`price`

Price per 33 cl. bottle of beer in Euros (1 Euro ≈ 10 Norwegian Crowns)

`country`

Country of production for beer: Norway = 0, outside of Norway = 1

`alch_perc`

Alcohol level (in percent) in the Christmas beer

`quality`

Taste quality of the beer on a scale from 0 (tasteless) to 10 (perfect taste)

Appendix B
Soccer Data

Data documentation for the data `soccer`; a data set on the soccer players in the Norwegian top-tier soccer league (Eliteserien) in 2014/2015. The only variable pertaining to 2015 is total income; the remaining variables pertain to the 2014 season. Variable names are in **bold** typeface. N = 240.

inc_year

Total, yearly income in Euros for player in the 2015 season (1 Euro ≈ 10 Norwegian Crowns)

age

Age of player

pos

Player position: goalkeeper = 0, defender = 1, midfielder = 2, attacker = 3

match_tot

Number of matches played during career

match_ses

Number of matches played during the season

goals

Number of goals scored during the season

assist

Number of assists made during the season

national

Number of matches played for the national team during career

club_rank

Ranking of a player's club at the end of the season: 1 = winner, 2 = second place, 3 = third place, …, 16 = last place. A *lower* number thus means playing for a better-performing or higher-ranked club.

origin

Player origin: Norwegian player = 0, foreign player = 1

nation_dum (recoding of **national**: 0 = 0, > 0 = 1)

Played for the national team at all: no = 0, yes = 1

age_ord

Recoding of **age**: 18–20 = 1, 21–23 = 2, 24–26 = 3, 27–29 = 4, 30–32 = 5, 33–41 = 6

log_inc

Natural logarithm of **inc_year**

Appendix C
Student Exercise Data

Data documentation for the data student_exercise; a survey questionnaire data for a random sample of students attending a Norwegian university college in 2017. Variable names are in **bold** typeface. N = 644.

times_exer

In a typical week, how many times do you exercise?

hours_exer

In a typical week, how many hours do you exercise?

exer_most

What kind of exercise do you prefer (one answer!)? strength = 0, cardio = 1, both equally much = 2

fitness_cen

Are you currently a member of a fitness center? no = 0, yes = 1

sport_cub:

Are you currently a member of a sports club? no = 0, yes = 1

age

Your age (in years)?

gender

Your gender? female = 0, male = 1

km_away

How far from downtown do you reside (in number of kilometers)?

status

Your current status? single = 0, boyfriend/girlfriend = 1, cohabiting/married = 2

health

How is your general physical health? ok = 0, good = 1, very good =2

econ

How is your financial situation? not good = 0, ok = 1, good = 2

youth_exe

In your youth before you started studying, to what extent were you involved in sports requiring lots of physical exercising (to a very small extent = 1, to a very great extent = 10)?

Recodings of existing variables:

times_ex_gr

Grouping of **times_exer**: 0–1 time a week = 0, 2–3 times a week = 1, 4 times or more a week = 2

hours_ex_gr

Grouping of **hours_exer**: 0–1 hour a week = 0, >1 to 3 hours a week = 1, >3 to 5 hours a week = 2, >5 hours a week = 3

health_dum

Grouping of **health**: ok/good = 0, very good = 1

weight

A so-called weighting variable to create a 50:50 distribution on the **gender** variable; cf. Chapter 6

3 Associational Research Questions I

Bivariate Analysis

3.1 Introduction: The Association between Two Variables, x and y

Variable associations are the bread and butter of quantitative research in the social and behavioral sciences. In this respect, we typically have some theoretical reason for expecting that one variable, x, affects another variable, y, in some way.[1] We then use a statistics program to examine our beliefs. I repeat: Our beliefs about how x is associated with y make us scrutinize this relationship empirically. For this reason, Chapter 2 was mainly an appetizer to the main event of statistical analysis in research settings: the examination of statistical associations. Now, let the real games begin![2]

There are several kinds of statistical associations, but three go a long way: differences in proportions, differences in means, and correlations. Sections 3.2 to 3.4 look at these three types of associations, corresponding with three different types of statistical analysis techniques. Furthermore, and as the section headings imply, the choice of which technique to use depends on the measurement level of the x-variable and y-variable in question. The three statistical analysis techniques covered in Sections 3.2 to 3.4 – cross-tabulation, one–way ANOVA, and regression analysis – also serve as the foundation for Section 3.5: the bivariate analysis in which the y-variable is ordinal.

Sections 3.6 and 3.7 spin around the limitations of bivariate analysis and lay down the foundation for Chapter 4. Section 3.8 summarizes the chapter and lists the key learning points, whereas Section 3.9 shows the do–file and syntax–file commands necessary to produce the results mentioned in the chapter. Section 3.10 provides some exercises with solutions.

Note! Throughout Chapter 3, my comments regarding the various associational results pertain only to what happens *within* the data. I do not refer to what might happen (or not) outside of the data. The latter inference–topic is for Chapter 5.

3.2 A Categorical x and a Categorical y: Cross-Tabulation

Table 2.9 showed that 22 percent of our students were members of sports clubs and that 78 percent were not. Suppose previous research among older adults has shown that men tend to be sports club members more often than women. Based on this, we expect such a pattern to be the case among our students as well. We thus ask if the *proportion*

DOI: 10.4324/9781003252559-3

Table 3.1 Sports club membership by gender in the student exercise data. Cross-tabulation. N = 644.

Variable: member of sports club	Female student	Male student	Total
No	86%	65%	78%
Yes	14%	35%	22%

of sports club members varies among female and male students in our data. Performing the statistical technique called cross-tabulation sheds light on this research question, and Table 3.1 presents the results.

The column on the right tells us what we already knew; the new information is in the two columns on the left of the total column. We note that 35 percent of the male students are sports club members, whereas the similar proportion for the female students is 14 percent. That is, we find that male students more often than female students are members of sport clubs – in accordance with our expectation. More precisely, the table shows an association between gender and sports club membership because the gender-specific proportions vary. This association boils down to the 21-percentage points difference between male and female membership propensity: 35–14 = 21. Since the only other option in the table is not being a sports club member, it follows that the gender difference for this negative option must be 21 percentage points as well.

Now for a thought experiment. What if the membership proportions were similar for female and male students, yielding a percentage difference of zero? This would suggest *no* association between the two variables. Statistical analysis in research is often about distinguishing between associations and non-associations in this way, and we will get back to this distinction several times in the book.

The Stata-output yielding the results in Table 3.1 appears in Stata-output 3.1. SPSS-output 3.1 of course looks very similar. The outputs from most other statistics programs look very close to the Stata-outputs and SPSS-outputs in a visual sense throughout this chapter as well.

```
. tab sport_club gender, col

              |         gender
  sport_club |    female      male |     Total
-------------+----------------------+----------
          no |       356       148 |       504
             |     85.58     64.91 |     78.26
-------------+----------------------+----------
         yes |        60        80 |       140
             |     14.42     35.09 |     21.74
-------------+----------------------+----------
       Total |       416       228 |       644
             |    100.00    100.00 |    100.00
```

Stata-output 3.1 Stata-results producing Table 3.1.

sport_club * gender Crosstabulation

			gender		
			female	male	Total
sport_club	no	Count	356	148	504
		% within gender	85,6%	64,9%	78,3%
	yes	Count	60	80	140
		% within gender	14,4%	35,1%	21,7%
Total		Count	416	228	644
		% within gender	100,0%	100,0%	100,0%

SPSS-output 3.1 SPSS-results producing Table 3.1.

Cross-tabulations as a way of presenting an association between a categorical *x* and a categorical *y* is a common undertaking in the social and behavioral sciences. In the most basic terms, a cross-tabulation counts how many units belong in each of the cross-table's cells. There are four cells in Table 3.1; we disregard the column for the totals. We might want to display the results of a cross-tabulation in a graph, as in Figure 3.1. We call this figure a horizontal and stacked bar chart. SPSS uses the vertical version as default; cf. the syntax in Section 3.9.

We do not always have a reasoned expectation for a potential association between *x* and *y* as in the case above. Sometimes, out of sheer curiosity – that is, from an exploratory point of view in method speak – we want to find out if a categorical *x* and a categorical *y* are associated somehow. Table 2.10 showed 34 percent of the students preferred doing strength training, 23 percent preferred doing cardio training, and 43

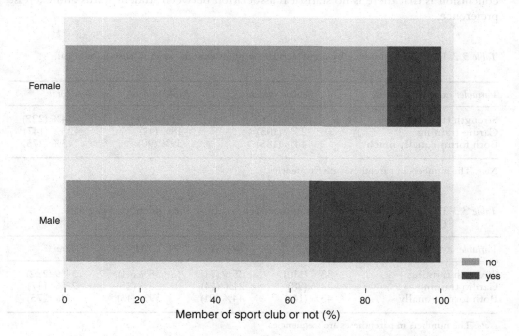

Figure 3.1 Graphical presentation of the cross-tabulation results in Table 3.1.

percent preferred doing both types of exercising equally. Do these preferences vary by gender? Table 3.2 presents a cross-tabulation answering this question. I have also added actual frequencies (i.e., the number of students) in each cell of the cross-table since many prefer this.

The horizontal gender comparisons for the two middle columns again concern us. We note in the uppermost row that male students tend to prefer strength training more often than female students. This gender difference between the two proportions is 12 percentage points. For cardio training, we find the opposite tendency: Female students tend to prefer this more often than male students, with a percentage points difference of seven. Finally, the category of doing equally has the most even gender distribution – 44 percent (male) versus 39 percent (female).

Summing up, there is some gender variation in exercise preferences because the percentage differences between the gender-specific proportions are higher than zero; we note the contour of a statistical association between the two variables. Again, what if the gender differences between the proportions were zero or thereabouts? Then we would be reluctant to claim that such an association existed.

A final example. Does the exercise preference vary by student status of being single, boy/girlfriend, or cohabiting/married? Table 3.3 sheds light on this research question by a cross-tabulation in the usual manner.

We have to make two horizontal comparisons for each preference in Table 3.3. Among the single students, we find that 33 percent prefer doing strength training. For the boyfriend or girlfriend group (BF/GF) the analogous number is 35 percent, whereas it is 38 percent for the married or cohabiting group (C/M). That is, the preference for doing strength training does not seem to vary much among the three groups. A similar pattern holds for the cardio preference and the both-forms preference. The tentative conclusion is that there is no statistical association between student status and exercise preference.

Table 3.2 Exercise preference by gender in the student exercise data. Cross-tabulation. $N = 644$.

Variable: exercise preference	Female student	Male student	Total
Strength training	30% (126)	42% (96)	34% (222)
Cardio training	25% (105)	18% (42)	23% (147)
Both forms equally much	44% (185)	39% (90)	43% (275)

Note. The numbers in parentheses are frequencies.

Table 3.3 Exercise preference according to student status in the student exercise data. Cross-tabulation. $N = 644$.

Variable: exercise preference	Single	BF/GF	C/M	Total
Strength training	33% (110)	35% (71)	38% (41)	34% (222)
Cardio training	24% (78)	21% (44)	23% (25)	23% (147)
Both forms equally	43% (141)	44% (91)	39% (43)	43% (275)

Note. The numbers in parentheses are frequencies.

We may in principle associate any two categorical variables in cross-tables like those shown above. But suppose your x (your variable x, that is!) and y have five categories each. This yields a cross-table with 25 cells: $5 \times 5 = 25$. Such a table is difficult to interpret and, more importantly, challenging to communicate to readers. For this reason, we often choose to regroup (*recode* in statistics lingo) a categorical variable with many categories into a variable having fewer categories before doing the cross-tabulation (more on this in Section 6.2). Now it is time for our second statistical analysis technique, covering the case of a categorical x and a continuous y. Section 3.3 is about one-way ANOVA.

3.3 A Categorical x and a Continuous y: ANOVA

We saw in Table 2.5 that the mean income among the soccer players was 87,367 Euros. Earlier research suggests that the earnings of athletes depended on many x-variables. For example, research has shown that players representing their national teams earn more than players not representing their national teams on average. (Playing for a national team signals better performances, which in turn tends to pay more.) We expect such a pattern among our Norwegian soccer players as well. We thus ask if the *mean* of income varies for national team players and players not representing their national teams. Applying the statistical technique called one-way ANOVA sheds light on this research question, and Table 3.4 presents the results.[3] Despite its complicated name, the ANOVA technique simply calculates and displays the mean of y for different subgroups in the data (to simplify a bit).

The bottom row shows what we already know for all the players; cf. Table 2.5. The main message of Table 3.4 is the huge mean difference in the two player groups' yearly income. The national team players earn 71,210 Euros more on average than the players not representing their national teams ($133,840 - 62,630 \approx 71,210$). This supports our expectations based on previous research.

Table 3.4 shows an association between player group and player income because the mean income is different for the two groups of players. The association boils down to the 71,210 Euros difference between the two groups. Now for a second thought experiment. What if the means were similar for the two groups of players, yielding a mean income difference of zero or thereabouts? Then we would claim there was no association between the two variables. The Stata-output responsible for the results of Table 3.4 appears in Stata-output 3.2.[4] The analogous SPSS-output is in SPSS-output 3.2.

Table 3.4 Mean of yearly income in Euros in 2015, in total and for national team players and players not representing their national teams in the soccer player data. One-way ANOVA.

Variable: yearly income in Euro	N =	Mean	SD
National team player: no	160	62,630	41,787
National team player: yes	80	133,840	86,025
Total	240	86,367	68,843

```
. oneway inc_year nation_dum, t
```

| | Summary of inc_year | | |
nation_dum	Mean	Std. Dev.	Freq.
no	62629.709	41787.042	160
yes	133840.44	86025.093	80
Total	86366.62	68843.27	240

Analysis of Variance

Source	SS	df	MS	F	Prob > F
Between groups	2.7045e+11	1	2.7045e+11	74.65	0.0000
Within groups	8.6226e+11	238	3.6230e+09		
Total	1.1327e+12	239	4.7394e+09		

Bartlett's test for equal variances: chi2(1) = 59.3339 Prob>chi2 = 0.000

Stata-output 3.2 Stata-results producing Table 3.4.

Descriptives

inc_year

	N	Mean	Std. Deviation	Std. Error	95% Confidence Interval for Mean		Minimum	Maximum
					Lower Bound	Upper Bound	Minimum	Maximum
no	160	62629,7094	41787,04187	3303,55573	56105,1993	69154,2194	5500,30	300364,20
yes	80	133840,4400	86025,09265	9617,89775	114696,4934	152984,3866	21808,30	458894,10
Total	240	86366,6196	68843,26951	4443,81394	77612,5754	95120,6638	5500,30	458894,10

ANOVA

inc_year

	Sum of Squares	df	Mean Square	F	Sig.
Between Groups	2,705E+11	1	2,705E+11	74,649	,000
Within Groups	8,623E+11	238	3622957776		
Total	1,133E+12	239			

SPSS-output 3.2 SPSS-results producing Table 3.4.

The comparison of subgroup means is a common way of finding out about two variables' potential association in the behavioral and social sciences. Again, we might want to present our results in a graph. Figure 3.2 illustrates this, and the vast difference in mean income between the two groups is apparent.

We often have an *a priori* expectation (i.e., before doing the analysis) about the association between *x* and *y*. Yet as before we sometimes do a statistical analysis just to find out if two variables are associated or not. Table 3.5 looks at the potential association between alcohol level and production location in the Christmas beer data. The table shows that the mean levels of alcohol are very similar for the two subgroups; 8.13 versus 8.29. That is, we find no association of importance (it is tempting to say substance!) between the two variables.

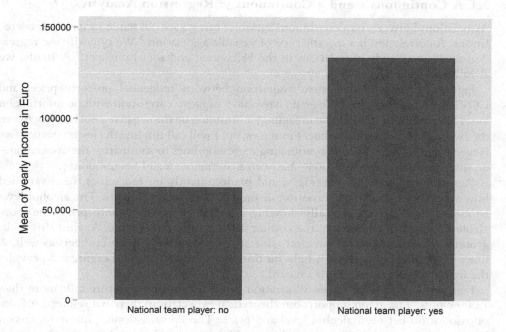

Figure 3.2 Graphical presentation of the ANOVA results in Table 3.4.

Table 3.5 Mean of alcohol level, in total and for two production locations in the Christmas beer data. One-way ANOVA.

Variable: alcohol level (percent)	N =	Mean	SD
Production location: Norway	47	8.13	1.42
Production location: outside of Norway	28	8.29	1.80
Total	75	8.19	1.57

Table 3.6 Mean of yearly income in Euros in 2015, in total and for four player positions in the soccer player data. One-way ANOVA.

Variable: yearly income in Euros	N =	Mean	SD
Goalkeeper	23	87,593	85,982
Defender	84	89,968	71,684
Midfielder	76	80,115	61,628
Attacker	57	88,900	67,383
Total	240	86,367	88,843

One-way ANOVA is not restricted to comparing the mean of y for two subgroups; we may compare as many subgroups as we like. Table 3.6 shows the mean income for the four player positions in our soccer data. Overall, there seems to be little income variation among the four positions. The two variables are not associated.

The time has come for our third statistical technique, relating two continuous variables. Section 3.4 is about regression analysis.

3.4 A Continuous *x* and a Continuous *y*: Regression Analysis

Associations between two continuous variables carry a special name in statistics: correlations. A correlation is a specific type of variable association.[5] We typically use regression analysis to study correlations in the behavioral and social sciences.[6] Actually, we already did so in Section 1.1.

Figure 1.1 showed a negative association between residential property prices and COVID-19 infection rates. We could have said a negative correlation with no information loss. I referred to the straight line in Figure 1.1 summing up the negative association between the two variables as the trend line. From now on, I will call this line the *linear regression line*. Regression analysis is all about estimating regression lines to summarize the association – or, more precisely, the correlation – between a continuous *x* and a continuous *y*.[7]

Research on alcoholic beverages, and predominantly on red wines, has examined how several factors explain variation in the prices of such products. The alcohol level is one of these factors, typically showing a positive correlation with price: the more alcohol content in a beverage, the costlier it tends to be on average. Against this background, we expect such a statistical association among our Christmas beers as well. A linear regression analysis sheds light on this research question, and Figure 3.3 provides the first graphical answer in this regard.

Figure 3.3 is perhaps a keen illustration of the adage that a picture tells more than 1,000 words. The graph supports our theoretical expectation; there is a *positive correlation* or association between alcohol level and price. The upward-sloping linear regression line from left to right suggests that beers with higher alcohol levels seem to be pricier on average than beers with lower alcohol levels.

Now we enter our thought experiment for the final time. What if the beers in Figure 3.3 were located in a way that yielded a *horizontal* regression line, implying that

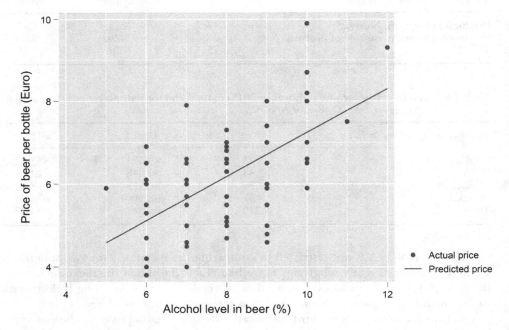

Figure 3.3 Scatterplot of correlation between alcohol level in beers and price of beers, with regression line.

the beers on average had the same price for every level of alcohol? Then we would say there was a zero correlation or a non-correlation between the two variables. That is, a *non*-horizontal regression line implies some form of correlation or association between x and y, whereas a horizontal regression suggests a zero-correlation or non-association.

Technically, a regression equation yields the regression line in Figure 3.3. In the Christmas beer case, this regression equation is

price of beer $= 1.94 + 0.53 \times$ alcohol level.

The number 1.94 is the spot where the regression line crosses the y-axis (i.e., the price-axis) if we extend it towards the left. The number 0.53 refers to the steepness of the regression line. That is, if we move one place to the right on the x-axis – from an alcohol level of six to seven, say – the bottle price is 0.53 Euros more expensive on average. Furthermore, by moving 2 percent levels to the right on the x-axis, the analogous price difference is 1.06 Euros and so on.

Another name for the number 1.94 is the *constant* or b_0. Furthermore, the number 0.53 is the *slope* or b_1. By replacing beer price with y and alcohol level with x, and suppressing the multiplication sign, we get the general regression equation

$y = b_0 + b_1 x$.

Figure 3.4 shows this general equation visually.[8] We note that the regression line crosses the y-axis at $x =$ zero; this is the constant/b_0. We also note that the slope/b_1 is the average change in y given a one-unit increase in x. And since the regression line is linear, it does not matter where we start on the x-axis when doing this one–unit–increase step. For the rest of this book, I will mostly use the term *regression coefficient* or b_1 when referring to the slope.

As it happens, statistics programs do not typically produce graphs (plots) as defaults when instructed to run a regression. That is, regression plots are mainly pedagogical devices to inform readers about what is going on behind the scenes. Stata-output 3.3 presents the results yielding the regression plot in Figure 3.3; SPSS-output 3.3 presents the analogous results in SPSS format.

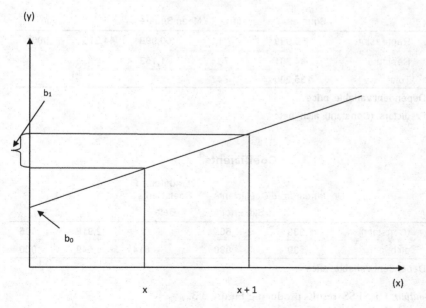

Figure 3.4 The regression line, the b_0 (i.e., constant), and the b_1 (i.e., regression coefficient/slope).

```
. reg price alch_perc

      Source |       SS           df       MS            Number of obs   =        75
-------------+------------------------------            F(1, 73)        =     44.22
       Model |  50.9992965            1  50.9992965     Prob > F        =    0.0000
    Residual |  84.2007035           73  1.15343429     R-squared       =    0.3772
-------------+------------------------------            Adj R-squared   =    0.3687
       Total |       135.2           74  1.82702703     Root MSE        =     1.074


-------------------------------------------------------------------------------------
       price |     Coef.   Std. Err.      t    P>|t|     [95% Conf. Interval]
-------------+-----------------------------------------------------------------------
   alch_perc |   .5302485   .0797432     6.65   0.000     .3713205    .6891765
       _cons |   1.939033   .6645055     2.92   0.005      .614675     3.26339
-------------------------------------------------------------------------------------
```

Stata-output 3.3 Stata-results producing Figure 3.3.

In Stata, the regression coefficient or b_1 appears under the heading Coef. for the variable alch_perc. (0.530 = 0.530.) Similarly, we find the constant or b_0 (1.939) under the same heading labelled as _cons. We will get back to other parts of the regression output later in the book.

Model Summary

Model	R	R Square	Adjusted R Square	Std. Error of the Estimate
1	,614[a]	,377	,369	1,07398

a. Predictors: (Constant), alch_perc

ANOVA[a]

Model		Sum of Squares	df	Mean Square	F	Sig.
1	Regression	50,999	1	50,999	44,215	,000[b]
	Residual	84,201	73	1,153		
	Total	135,200	74			

a. Dependent Variable: price

b. Predictors: (Constant), alch_perc

Coefficients[a]

Model		Unstandardized Coefficients		Standardized Coefficients	t	Sig.
		B	Std. Error	Beta		
1	(Constant)	1,939	,665		2,918	,005
	alch_perc	,530	,080	,614	6,649	,000

a. Dependent Variable: price

SPSS-output 3.3 SPSS-results producing Figure 3.3.

In SPSS, the regression coefficient appears under the heading B for the variable `alch_perc`. (0.530 = 0.530.) Similarly, we find the constant (1.939) under the same heading labeled as (`Constant`).

We may say the regression coefficient measures the size of the correlation between two continuous variables. Yet a more typical measure in this regard is the Pearson correlation coefficient – or r. The r for the correlation between alcohol level and beer price is 0.6142.[9] Possible values for r lie in the −1 to +1 range, where positive values imply a positive correlation and negative values suggest a negative correlation. The value of zero suggests no (linear) correlation.

Let us look at another correlation in the soccer data. Intuition and prior research suggest that players with more experience earn more than players with lesser experience. Figure 3.5 shows this correlation in a scatterplot that also includes the regression line, and Stata-output 3.4 presents the results generating this regression line. There is no need to present the analogous SPSS-output; it looks very similar to Stata's as we just saw in the previous example.

The plot – but mostly the regression line[10] – supports our common-sense notion; more experienced players tend to earn more than less experienced players: A player with, say, 80 matches in his career earns 248 Euros more on average than a player with 79 matches in his career; cf. Stata-output 3.4. This amounts to 12,400 Euros a year for a 50-match difference ($248 \times 50 = 12,400$).

Now, remember the regression coefficient's literal interpretation: the average change in y given a one-unit increase in x. It might thus be tempting to claim that playing one more match increases the income with 248 Euros, but this interpretation is inaccurate.

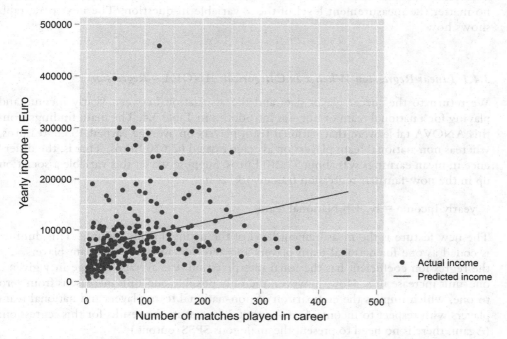

Figure 3.5 Scatterplot of correlation between number of matches played in career and yearly income, with regression line.

The regression compares players with unequal match experience with respect to income at one point in time, that is, in a static manner. This regression has little to offer on any dynamic income-changes among the players.

```
. reg inc_year match_tot

      Source |       SS          df        MS          Number of obs  =      240
-------------+----------------------------------        F(1, 238)      =    20.76
       Model |   9.0878e+10        1    9.0878e+10      Prob > F       =   0.0000
    Residual |   1.0418e+12      238    4.3775e+09      R-squared      =   0.0802
-------------+----------------------------------        Adj R-squared  =   0.0764
       Total |   1.1327e+12      239    4.7394e+09      Root MSE       =    66162


------------------------------------------------------------------------------
    inc_year |      Coef.   Std. Err.       t    P>|t|     [95% Conf. Interval]
-------------+----------------------------------------------------------------
   match_tot |   248.2282   54.47967     4.56   0.000     140.9042    355.5521
       _cons |    65166.9   6315.689    10.32   0.000     52725.11    77608.69
------------------------------------------------------------------------------
```

Stata-output 3.4 Results producing Figure 3.5.

Regression is mainly about finding the overall trend in a correlation between x and y. The regression line is the key in this respect, whereas the individual data points (i.e., the dots/matches in Figure 3.5) are of less importance. It follows that the regression equation and the regression coefficient/b_1 get the lion's share of attention. This emphasis on b_1 has one important consequence: The measurement level of the x-variable becomes more or less irrelevant. That is, we may do regression analysis no matter the measurement level of the x-variable in question. The next paragraph shows how.

3.4.1 Linear Regression When x Is Categorical: ANOVA ≈ Regression

We return to the soccer player data and the association between yearly income and playing for a national team or not – as introduced in Table 3.4. The main findings from this ANOVA table were that national team players on average earned 133,840 Euros, whereas non-national team players on average earned 62,630 Euros. That is, the difference in mean earnings was about 71,210 Euros. Suppose we set this variable association up in the now-familiar regression framework, as in

yearly income = b_0 + b_1National team.

The new feature is the measurement level of the national team variable. This dummy is coded as one for national team players and zero for non-national team players. Yet the regression coefficient has the usual interpretation: the average change in y given a one-unit increase in x. Now, however, the only possible one-unit increase is from zero to one, which implies the comparison of non-national team players and national team players with respect to income. Stata-output 3.5 presents the results for this regression. (Again, there is no need to present the analogous SPSS-output.)

```
. reg inc_year i.nation_dum

      Source |       SS           df       MS        Number of obs   =        240
-------------+----------------------------------     F(1, 238)       =      74.65
       Model |   2.7045e+11         1   2.7045e+11   Prob > F        =     0.0000
    Residual |   8.6226e+11       238   3.6230e+09   R-squared       =     0.2388
-------------+----------------------------------     Adj R-squared   =     0.2356
       Total |   1.1327e+12       239   4.7394e+09   Root MSE        =      60191


--------------------------------------------------------------------------------
    inc_year |      Coef.   Std. Err.       t    P>|t|     [95% Conf. Interval]
-------------+------------------------------------------------------------------
  nation_dum |
         yes |   71210.73   8241.994     8.64    0.000     54974.16    87447.31
       _cons |   62629.71   4758.517    13.16    0.000     53255.52     72003.9
--------------------------------------------------------------------------------
```

Stata-output 3.5 Results for regression of yearly income by national team dummy in the soccer player data.

The literal interpretation is that when the national team dummy increases by one unit, that is, from zero to one, the income increases by 71,210 Euros. More correctly and more in step with reality, we note that national team players on average earn 71,210 Euros more than non–national team players – just as we did for the ANOVA in Table 3.4. (Do not pay attention to the correct rounding for the regression coefficient to 71,211.) What does the constant (_cons) mean here? Plugging the x-value for the non–national team players, that is, zero, into the regression equation, we get

$$\text{yearly income} = 62{,}630 + 71{,}210 \times 0 \rightarrow \text{yearly income} + 0 = 62{,}630.$$

The constant thus refers to the mean of yearly income for the non–national team players.[11] To obtain the mean of yearly income for the national team players, we plug the x-value for these players, that is, one, into the regression equation. We then get

$$\text{yearly income} = 62{,}630 + 71{,}210 \times 1 \rightarrow \text{yearly income} = 62{,}630 + 71{,}210 = 133{,}840.$$

These calculations show that one-way ANOVA and bivariate regression with a dummy are the same. The two techniques differ only with respect to the presentation of results: ANOVA reports the mean of y for the two subgroups directly. Regression reports similar information indirectly: the mean of y for one subgroup (i.e., the constant), and how much the other subgroup differs from the constant regarding the mean of y. This difference is the regression coefficient.

The results in this subsection have one important implication from an applied research perspective. Regression subsumes ANOVA and thus makes the latter redundant in many real-life applications. For this reason, I will mostly focus on regression rather than ANOVA from here on in the book.[12]

Coming up next is a regression in which the categorical x-variable has four categories. We saw a similar ANOVA style analysis in Table 3.6, that is, the mean of yearly income for the four soccer player positions. Stata-output 3.6 presents the results. (Again, there is no need to present the analogous SPSS-output.)

```
. reg inc_year i.pos

      Source |       SS           df       MS        Number of obs   =        240
-------------+-----------------------------------    F(3, 236)       =       0.31
       Model |   4.4607e+09        3  1.4869e+09     Prob > F        =     0.8174
    Residual |   1.1283e+12      236  4.7807e+09     R-squared       =     0.0039
-------------+-----------------------------------    Adj R-squared   =    -0.0087
       Total |   1.1327e+12      239  4.7394e+09     Root MSE        =      69143


------------------------------------------------------------------------------
    inc_year |      Coef.   Std. Err.      t    P>|t|     [95% Conf. Interval]
-------------+----------------------------------------------------------------
         pos |
     defense |   2375.894   16271.81     0.15   0.884    -29680.67    34432.45
    midfield |  -7477.857   16454.87    -0.45   0.650    -39895.05    24939.34
      attack |   1307.245   17080.13     0.08   0.939    -32341.76    34956.25
             |
       _cons |   87592.57   14417.29     6.08   0.000     59189.54    115995.6
------------------------------------------------------------------------------
```

Stata-output 3.6 Results for regression of yearly income by the categorical player position variable in the soccer player data.

The key to interpret the output is noticing the player position *not* showing up: The goalkeepers. The goalies earn 87,593 on average, as suggested by the constant. We compare the remaining player positions to this reference. For example, the defenders make 2,376 Euros more than the goalies on average. Furthermore, the midfielders make 7,478 Euros less than the goalies; cf. the negative sign of the coefficient. Still, the main message from Stata-output 3.6 echoes the ANOVA results of Table 3.6. There is little income variation among the four player positions.

3.4.2 Residential Property Prices and COVID-19 Spread Revisited

We close the regression introduction by reporting the coefficient and constant for the property price variable in the opening figure of the book, that is, Figure 1.1. Stata-output 3.7 displays the results. Again, there is no need to present the 100 percent analogous SPSS-output.

```
. reg infect_rate price_sq_m

      Source |       SS           df       MS        Number of obs   =         15
-------------+-----------------------------------    F(1, 13)        =      10.99
       Model |   93608.4181        1  93608.4181     Prob > F        =     0.0056
    Residual |   110682.611       13  8514.04702     R-squared       =     0.4582
-------------+-----------------------------------    Adj R-squared   =     0.4165
       Total |   204291.029       14  14592.2164     Root MSE        =     92.272


------------------------------------------------------------------------------
 infect_rate |      Coef.   Std. Err.      t    P>|t|     [95% Conf. Interval]
-------------+----------------------------------------------------------------
   price_sq_m |  -5.328929   1.607128    -3.32   0.006    -8.800918    -1.85694
       _cons |   677.2847   111.2654     6.09   0.000     436.9104    917.6591
------------------------------------------------------------------------------
```

Stata-output 3.7 Results producing Figure 1.1.

We note the expected negative regression coefficient, suggesting a downward–sloping regression line going from left to right.[13]

3.5 An Ordinal y and Bivariate Analysis

The statistical techniques covered thus far – the cross-table, the ANOVA, and the regression – are all alternatives at the outset for an ordinal y-variable. What often determines which technique to use is the number of ordered categories for the y in question. The student data includes a question about physical health; cf. Table 2.12. The frequency distribution for this variable is ok health = 34 percent, good health = 49 percent, and very good health = 17. We expect a positive association between exercising and health based on prior studies; we expect students exercising often to report better health than students exercising less often. The data also contain an exercise frequency variable (`times_ex_gr`). The frequency distribution for this variable is zero to one time per week = 15 percent, two to three times per week = 47 percent, and four times or more per week = 39 percent. (There is no need to show these results in a table.) That is, the y-variable (health) and the x-variable (exercise frequency) are both ordinals having three categories each. It is always safe to do a cross-tabulation in such cases. Table 3.7 presents the results.

The column on the right repeats the overall answer distribution for the health variable, and we begin the comparisons for the bottom row. Among students exercising zero to one time per week, 7 percent report very good health. Similarly, among students exercising two to three times per week, 9 percent answer very good health. Finally, for students exercising four times per week or more, 31 percent report very good health. We find the opposite pattern for the ok health group in the top row: Students exercising infrequently have the highest probabilities of reporting ok health, whereas students exercising often have the least probabilities of reporting ok health. (The tendency for the good health category in the middle row is akin to the very good category in the bottom row.) Summing up, the results support a "positive association" between exercise frequency and health: Compared with students exercising infrequently, students exercising often tend to report having very good physical health more often.

Cross-tabulation is always a possibility for associating two ordinal variables having few categories. Yet such an analysis neglects the ordered nature of the variables. If we want one summary measure for the association between two ordinal variables – akin to the regression coefficient or Pearson's r for two continuous variables – we ask the statistics program for Kendall's Tau ordinal correlation coefficient. This coefficient is

Table 3.7 Physical health assessment by exercise frequency in the student exercise data. Cross-tabulation. $N = 644$.

Variable: physical health	0–1 p/w	2–3 p/w	4 or more p/w	Total
Ok	59% (57)	40% (120)	16% (40)	34% (217)
Good	33% (32)	51% (152)	53% (131)	49% (315)
Very good	7% (7)	9% (28)	31% (77)	17% (112)

Note. The numbers in parentheses are frequencies.

Table 3.8 Physical health assessment by gender in the student exercise data. Cross-tabulation. N = 644.

Variable: physical health	Female student	Male student	Total
Ok	34% (142)	33% (75)	34% (217)
Good	51% (213)	45% (102)	49% (315)
Very good	15% (61)	22% (51)	17% (112)

Note. The numbers in parentheses are frequencies.

0.34 in the present case. A limitation with this summary measure compared to the very transparent cross-table is that the group differences in health get somewhat lost.

Suppose we want to find out if there is an association between the age of the students and their physical health level. In this case, a cross-tabulation will not cut it because age is a continuous variable. Imagine three health level cells for 19-year old, three for 20-year old, and three for 21-year old students, and so on that will result in a vast cross-table! A way to solve this is first to recode the age variable into an ordered age variable with categories such as 19 to 21 years, 22 to 24 years, 25 to 27 years, and so on (see Section 6.2 for more on recoding).[14] That done, we may associate the new and ordered age variable to the exercise frequency variable in a cross-tabulation or by employing an ordinal correlation analysis.[15]

Table 3.8 examines the potential association between gender and physical health using a cross-tabulation. The results show that male students are slightly more likely than female students to report very good health, whereas it is the other way round for good health. In this case, however, doing ordinal correlation analysis is not a viable option because the x-variable is categorical and not ordinal.

3.5.1 An Ordinal y Having Many Categories

When an ordinal y-variable has many categories, say about six or more, it is common and most often uncontroversial to treat it as continuous when using it in a bivariate analysis.[16] The next example illustrates. The ordinal taste variable in the Christmas beer data has 11 categories ranging from zero (terrible) to ten (perfect). Figure 3.6 examines if the alcohol level is associated with the quality scores of the beers using regression analysis.

The new feature of this regression is that the quality scores have a limited range of values. The more important upshot is that the usual linear regression usually works fine for ordinal y-variables having many categories. The regression coefficient in question, which by now we need not present, is 1.01. This suggests that a beer with, say, a 7-percent alcohol level on average gets a 1.01 points better quality score than a beer with an alcohol level of 6 percent.

A final example concerns the y-variable youth sports involvement in the student data. This y reads, "In your youth before you started studying, to what extent were you involved in sports requiring lots of physical exercise (to a very small extent = 1; to a very great extent = 10)?" The mean is 6.56 on this one-to-ten scale, as we saw in Table 2.11. Below we examine if this mean value varies by gender. The regression results appear in Stata-output 3.8. Again, there is no need to present the analogous SPSS-output.

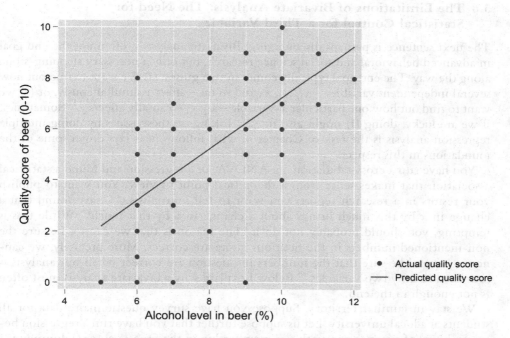

Figure 3.6 Scatterplot of correlation between alcohol level in beers and quality of beers, with regression line.

```
. reg youth_exe i.gender

      Source |       SS           df       MS         Number of obs   =       644
-------------+----------------------------------       F(1, 642)       =     29.94
       Model |  241.217518         1   241.217518      Prob > F        =    0.0000
    Residual |  5173.17223       642   8.05790068      R-squared       =    0.0446
-------------+----------------------------------       Adj R-squared   =    0.0431
       Total |  5414.38975       643   8.42051283      Root MSE        =    2.8386

------------------------------------------------------------------------------
   youth_exe |      Coef.   Std. Err.      t    P>|t|     [95% Conf. Interval]
-------------+----------------------------------------------------------------
      gender |
        male |   1.279774   .2339052     5.47   0.000     .8204624    1.739086
       _cons |   6.110577    .139176    43.91   0.000     5.837282    6.383872
------------------------------------------------------------------------------
```

Stata-output 3.8 Results for regression of youth sports involvement by gender in the student exercise data.

The literal interpretation is that the youth sports variable increases by 1.28 points when the gender dummy increases by one unit. More correctly, male students score 1.28 points higher on the youth sports variable than female students on average. The female students score 6.11 for this variable on average, as indicated by the constant. I could have said more on ordinal bivariate analysis, but I will stop here. Literature for further reading on this topic appears at the end of the chapter.

3.6 The Limitations of Bivariate Analysis: The Need for Statistical Control for a Third Variable

The next sentence is perhaps discouraging. Bivariate analysis is often not the end goal in advanced behavioral and social science research, but only a necessary stepping-stone along the way. The end goal typically comes in two guises: (1) We want to find out how several independent variables – x_1, x_2, x_3 and so on – affect y simultaneously, or (2) we want to find out how one particular x-variable – say x_1 – causally affects y.[17] Sometimes, if we are lucky, doing (1) might also fix (2). Taking on these issues by doing multiple regression analysis is the task of Chapter 4; what follows next lays down some of the foundations in this respect.

You have run a cross-tabulation, an ANOVA, or a regression and found a statistical association that makes sense from a theoretical point of view. You want to publish your results in a research report; you want to tell journalists, "I have found that a change in x by this much brings about a change in y by that much!" While this is tempting, you should probably not do it. The reason is that we cannot be sure the non-mentioned numbers in the fictitious quote are correct. More precisely, we cannot in general be sure that the numbers in question are correct when our analysis is restricted to only two variables.[18] Below I explain why a bivariate analysis most often is not enough in the end.

We stay in familiar territory. Suppose you have survey questionnaire data for all students at a local university. Let us suppose further that you have run a regression between hours of exercise per week and membership in the fitness center (a dummy; yes = 1, no = 0) and found a regression coefficient of two: Fitness center members exercise two more hours per week on average than non-members. This is as expected. The key question concerns the interpretation of this bivariate association, that is, whether it is reasonable or not to think of the two-hour regression coefficient as a causal effect. In this regard, the idea that buying a subscription to a fitness center will cause students to exercise more hours per week is plausible. Yet does buying such a subscription *in itself* make students exercise *exactly* two more hours per week on average? That is to say, is the effect causal? This appears much less plausible. The upcoming story tells you why. In it, you play the role of a sales representative for a fitness center.

You run into fellow students Liza and Marie, both presently not members of fitness centers, and ask them to buy a center subscription. You advertise that such a subscription increases exercising by two hours per week "according to one recent study." Both sign up! Fast-forward two months, and to when you meet Liza and Marie again and ask them about their present exercise habits. Liza, halfway happy, says she exercises for the same number of hours as she did before buying the subscription, but that most of the exercising now happens inside the center. Marie, in contrast, is delighted. Since signing, she has increased her exercise by four hours per week. The point now is not that the advertisement got it right on average (i.e., $(0 + 4)/2 = 2$). The point is that the effect of member subscription is *different* for both Liza and Marie because they differ regarding a potentially long list of x-variables causing variation in exercise hours. It is, for example, not a stretch to think of Liza as very motivated when it comes to exercising and, conversely, to imagine Marie as someone struggling with the motivation to exercise. In short: Liza's zero-hour effect and Marie's four-hour effect might have everything to do with their difference in exercise motivation *before* signing up for the fitness center subscription.[19]

To find the causal effect of how a fitness center subscription affects later exercise hours, we should ideally compare Liza with a woman having a similarly high level of exercise motivation who did not buy such a subscription. Analogously, we should ideally compare Marie with a woman similar to her when it comes to the low level of exercise motivation who did not buy a subscription. And so on for two Carries, two Jonis, and two Ericas, etc. Can we do this? The short answer is yes. A longer one is that we may – and often should! – include exercise motivation as a third x-variable in our analysis and carry out such a like-for-like comparison. The next paragraph sheds more light on this in a preliminary fashion.

Our imagined bivariate regression compares fitness center members and non-members as they are in a metaphorical sense; it does not consider the many differences between the two groups likely to cause a difference in exercise hours when calculating the overall mean group difference. In a manner of speaking, bivariate regression thus compares apples with oranges. Multiple regression, in contrast, explicitly accounts for these "many differences" – that is, a difference in motivation – when calculating the mean group difference in question. That is, multiple regression compares apples with rather similar apples. In our imagined case, the multiple regression analysis finds the effect of fitness center membership on exercise hours while holding exercise motivation constant at some fixed level. This *holding-constant* ability is the key competence of multiple regression, making it an indispensable tool in the quest to unravel causal effects in the social and behavioral sciences. Another name for this ability is statistical control: We find the effect of x_1 (fitness center membership) on y (hours of exercise) while *statistically controlling* for x_2 (exercise motivation), x_3, x_4, and so on.

In research, we often want to find the causal effect of x on y. Other terms for the causal effect are the *correct* effect, the *unbiased* effect, or the *unique* effect. You might have heard that correlation does not imply causation. In practice, this often means that a bivariate analysis does not say much about the causal effect of x on y for observational data. It is actually a bit more complicated, but that is the takeaway message.[20] Like-for-like comparisons are essential for quantitative research having causal ambitions, and the holding-constant principle briefly described above is one solution for it. I will explain the particulars in this regard in more depth in Chapter 4. The next short Section brings up another way out: the experimental control method.

3.7 Experimental Control for a Third (and Fourth) Variable

Imagine 100 women sharing the same disease entering a hospital reception. They have all signed up for a study on how a new medication will affect the illness they suffer from. The first woman in line, patient one, walks up to the desk and signs herself in. A nurse makes a metaphorical coin toss to provide patient one with a head or a tail. This metaphorical procedure repeats itself 99 more times for the remaining patients. The result of the process given a fair coin is an approximately 50:50 distribution of heads and tails. The formal term for this coin-toss procedure is *randomization*, and it ensures that the patient group receiving heads is similar to the patient group getting tails on average. More precisely, the coin toss ensures that any systematic differences between the two groups of patients before they enter the clinical trial (as they call it on Grey's) are the results of pure, random chance. Formally, we call the study type in question a Randomized Controlled Trial (RCT). Let us suppose that the heads-group gets the real medicine, and the tails-group gets the placebo (i.e., fake medicine). Note that neither the doctors nor the patients know who gets what; this type of trial is called "double blind."

Fast-forward two months to the diagnostic test for the disease in question. This test is the y-variable with values from, say, zero to ten. The medicine versus placebo variable is a dummy in the regression framework, where real medicine is coded one and placebo zero. The regression coefficient is thus the average difference in the diagnostic test between the medicine group and the placebo group (i.e., the constant). Note that the medicine or placebo → diagnostic test setup is analogous to the fitness center member or non-member → hours of exercise setup from a statistical viewpoint. Yet there is one vital difference: For the association between fitness center membership and exercise hours, we have few reasons to believe that the regression coefficient reflects the causal relationship between the two variables. For the RCT, in contrast, we have every reason to think the regression coefficient for the medicine or placebo dummy is the causal effect of the medicine on the diagnostic test! The next paragraph explains why.[21]

The keyword in the explanation is *randomization*. Randomization is the vital element in RCTs and all forms of experimental research designs. The problem in Section 3.6 was that members and non-members of a fitness center were different in many ways (e.g., exercise motivation) which might account for the mean difference in exercise hours. The "magical" feature of randomization is that it controls for all such alternative causes by the experimental design in itself. Moreover, randomization accounts for the alternative causes we may think of in advance of a study as well as those we forget to think of! In contrast, and as Chapter 4 will show in detail, statistical control only accounts for the alternative causes we enter into our multiple regression analysis.

The causal inference in RCTs/experimental designs has a Sherlock Holmes flavor: "Once we have eliminated the impossible, whatever remains, however improbable, must be the truth!"[22] Because randomization controls for all other possible causes, the cause that remains — that is, the medical treatment — must be the true cause no matter how unlikely it might sound.

Experimental designs/RCTs have become more popular in behavioral and social sciences of late. The reason is the straightforward causal inference if everything goes as planned in the experiment, which, of course, may or may not happen. That said, experiments are not always feasible in research having to do directly with people for ethical, political, or economic reasons. The statistical control approach in Chapter 4 is therefore oftentimes the only option we have. Finally, there are cases in which statistical control also might improve RCTs and experiments, and I return to this topic in Section 4.6.

3.8 Chapter Summary, Key Learning Points, and Further Reading

This chapter has been about analyzing associations between two variables, x and y, where we expect or assume the former to have a statistical effect on the latter: $x \rightarrow y$. Some key learning points follow below:

- We typically use three statistical analysis techniques to associate x and y: cross-tabulation, one-way ANOVA, and regression analysis.
- When to use the three techniques depends on what we assume is x (i.e., the independent variable or "cause"), what we assume is y (i.e., the dependent variable or "effect"), and the measurement level of the x-variable and y-variable in question.

- Regression analysis subsumes ANOVA in the bivariate case, making one-way ANOVA more or less redundant in practice. Yet ANOVA might be preferred as a personal choice or as a requirement in specific research fields.
- When a statistical association for observational data involves only two variables, x and y, we can never be sure to identify the causal or unique effect of x on y.
- We have two options when the aim is to identify x's causal effect on y: the statistical control procedure (which is imperfect) and the experimental control procedure (which is better). More on both procedures in Chapter 4.

The books mentioned in the further reading paragraph in Section 2.8 also apply to this chapter, but I would like to offer two more: Pearl and Mackenzie (2018) and Rosenbaum (2017).

3.9 Do-Files in Stata and Syntax-Files in SPSS

Make sure you have read Section 2.9 before taking on this section. I present the commands as they should appear in do-files (Stata) or syntax-files (SPSS), but I present them in plain text "outside" of such files to save space. I also add some comments to the various commands on occasion. I assume throughout that the "correct" data set is in memory to avoid unnecessary repetition.

3.9.1 Stata-Commands in Do-Files

Table 3.1/Stata-output 3.1

```
tab sport_club gender, col
```

The command above puts the y-variable (sport_club) in rows and the x-variable (gender) in columns, which I personally prefer. It also makes the percentages amount to a 100 vertically (col). You could turn this table one its head (rows → columns; columns → rows), for which you should use the command:

```
tab gender sport_club, row.
```

For new cross-tabulations, replace sport_club with a new variable name and gender with a new variable name.

Figure 3.1

```
catplot sport_club gender, percent(gender) asyvars stack
```

If you did not download the command catplot in Chapter 2, you must do it now; see Section 2.9.

Table 3.2

```
tab exer_most gender, col
```

Table 3.3

```
tab exer_most status, col
```

Table 3.4/Stata-output 3.2

```
oneway inc_year nation_dum, t
```

For new ANOVAs, replace `inc_year` with a new variable name and `nation_dum` with a new variable name.

Figure 3.2

```
graph bar (mean) inc_year, over(nation_dum)
```

Table 3.5

```
oneway alch_perc country, t
```

Table 3.6

```
oneway inc_year pos, t
```

Figure 3.3

```
twoway (scatter price alch_perc) (lfit price alch_perc)
```

For new but similar figures, replace `price` with a new variable name and `alch_perc` with a new variable name.

Stata-output 3.3

```
reg price alch_perc
```

For new regressions, replace `price` with a new variable name and `alch_perc` with a new variable name.
To obtain the correlation coefficient, *r*, between the variables in Stata-output 3.3, use the command:

```
corr price alch_perc
```

Figure 3.5

```
twoway (scatter inc_year match_tot) (lfit inc_year match_tot)
```

Stata-output 3.4

```
reg inc_year match_tot
```

Stata-output 3.5

```
reg inc_year i.nation_dum
```

The prefix `i.` in front of a categorical *x*-variable tells Stata to display the relevant label(s) for the *x*-variable of interest, that is, the category "yes" for nation_dum in this case.

Stata–output 3.6

```
reg inc_year i.pos
```

Stata–output 3.7

```
reg infect_rate price_sq_m
```

Table 3.7

```
tab health times_ex_gr, col
```

The command to get the ordinal correlation coefficient is:

```
ktau health times_ex_gr
```

Note that the value of 0.34 refers to Kendall's so-called Tau-b.

Table 3.8

```
tab health times_gender, col
```

Figure 3.6

```
twoway (scatter quality alch_perc) (lfit quality alch_perc)
```

Stata output 3.8

```
reg youth_exe i.gender
```

3.9.2 SPSS–Commands in Syntax–Files

As mentioned in Section 2.9, I use the paste-from-interactive-mode option to get the statistical commands into SPSS syntaxes. Once there, the copy-paste-replace-variable-names procedure saves much typing.

Table 3.1/SPSS–output 3.1

```
CROSSTABS
  /TABLES=sport_club BY gender
  /FORMAT=AVALUE TABLES
  /CELLS=COUNT COLUMN
  /COUNT ROUND CELL.
```

The command above puts the *y*-variable (sport_club) in rows and the *x*-variable (gender) in columns, which I personally prefer. It also makes the percentages amount to a 100 vertically (COLUMN). You could turn this table one its head (rows → columns; columns → rows), for which you should use the command:

```
CROSSTABS
  /TABLES=gender BY sport_club
  /FORMAT=AVALUE TABLES
```

```
/CELLS=COUNT ROW
/COUNT ROUND CELL.
```

For new cross-tabulations, replace sport_club with a new variable name and gender with a new variable name.

Figure 3.1 (vertically)

```
GGRAPH
  /GRAPHDATASET NAME="graphdataset" VARIABLES=gender COUNT()
[name="COUNT"] sport_club
    MISSING=LISTWISE REPORTMISSING=NO
  /GRAPHSPEC SOURCE=INLINE.
BEGIN GPL
  SOURCE: s=userSource(id("graphdataset"))
  DATA: gender=col(source(s), name("gender"), unit.category())
  DATA: COUNT=col(source(s), name("COUNT"))
  DATA: sport_club=col(source(s), name("sport_club"), unit.
category())
  GUIDE: axis(dim(1), label("gender"))
  GUIDE: axis(dim(2), label("Percent"))
  GUIDE: legend(aesthetic(aesthetic.color.interior),
label("sport_club"))
  GUIDE: text.title(label("Stacked Bar Percent of gender by
sport_club"))
  SCALE: cat(dim(1), include("0", "1"))
  SCALE: linear(dim(2), include(0))
  SCALE: cat(aesthetic(aesthetic.color.interior), include("0", "1"))
  ELEMENT: interval.stack(position(summary.percent(gender*COUNT,
base.coordinate(dim(1)))),
    color.interior(sport_club), shape.interior(shape.square))
END GPL.
```

Table 3.2

```
CROSSTABS
  /TABLES=exer_most BY gender
  /FORMAT=AVALUE TABLES
  /CELLS=COUNT COLUMN
  /COUNT ROUND CELL.
```

Table 3.3

```
CROSSTABS
  /TABLES=exer_most BY status
  /FORMAT=AVALUE TABLES
  /CELLS=COUNT COLUMN
  /COUNT ROUND CELL.
```

Table 3.4/SPSS–output 3.2

```
ONEWAY inc_year BY nation_dum
  /STATISTICS DESCRIPTIVES
  /MISSING ANALYSIS.
```

For new ANOVAs, replace inc_year with a new variable name and nation_dum with a new variable name.

Figure 3.2

```
GGRAPH
  /GRAPHDATASET NAME="graphdataset" VARIABLES=nation_dum MEAN(inc_
year)[name="MEAN_inc_year"]
    MISSING=LISTWISE REPORTMISSING=NO
  /GRAPHSPEC SOURCE=INLINE.
BEGIN GPL
  SOURCE: s=userSource(id("graphdataset"))
  DATA: nation_dum=col(source(s), name("nation_dum"), unit.
category())
  DATA: MEAN_inc_year=col(source(s), name("MEAN_inc_year"))
  GUIDE: axis(dim(1), label("nation_dum"))
  GUIDE: axis(dim(2), label("Mean inc_year"))
  GUIDE: text.title(label("Simple Bar Mean of inc_year by
nation_dum"))
  SCALE: cat(dim(1), include("0", "1"))
  SCALE: linear(dim(2), include(0))
  ELEMENT: interval(position(nation_dum*MEAN_inc_year), shape.
interior(shape.square))
END GPL.
```

Table 3.5

```
ONEWAY alch_perc BY country
  /STATISTICS DESCRIPTIVES
  /MISSING ANALYSIS.
```

Table 3.6

```
ONEWAY inc_year BY pos
  /STATISTICS DESCRIPTIVES
  /MISSING ANALYSIS.
```

Figure 3.3

```
GGRAPH
  /GRAPHDATASET NAME="graphdataset" VARIABLES=alch_perc price MISS-
ING=LISTWISE REPORTMISSING=NO
  /GRAPHSPEC SOURCE=INLINE
  /FITLINE TOTAL=YES.
```

```
BEGIN GPL
  SOURCE: s=userSource(id("graphdataset"))
  DATA: alch_perc=col(source(s), name("alch_perc"))
  DATA: price=col(source(s), name("price"))
  GUIDE: axis(dim(1), label("alch_perc"))
  GUIDE: axis(dim(2), label("price"))
  GUIDE: text.title(label("Simple Scatter with Fit Line of price by
alch_perc"))
  ELEMENT: point(position(alch_perc*price))
END GPL.
```

For new but similar figures, replace price with a new variable name and alch_perc with a new variable name.

SPSS-output 3.3

```
REGRESSION
  /DESCRIPTIVES MEAN STDDEV CORR SIG N
  /MISSING LISTWISE
  /STATISTICS COEFF OUTS R ANOVA
  /CRITERIA=PIN(.05) POUT(.10)
  /NOORIGIN
  /DEPENDENT price
  /METHOD=ENTER alch_perc.
```

For new regressions, replace price with a new variable name and alch_perc with a new variable name.

To obtain the correlation coefficient, *r*, between the variables in SPSS-output 3.3, use the command:

```
CORRELATIONS
  /VARIABLES=price alch_perc
  /PRINT=TWOTAIL NOSIG
  /MISSING=PAIRWISE.
```

Figure 3.5

```
GGRAPH
  /GRAPHDATASET NAME="graphdataset" VARIABLES=match_tot inc_year
MISSING=LISTWISE REPORTMISSING=NO
  /GRAPHSPEC SOURCE=INLINE
  /FITLINE TOTAL=YES.
BEGIN GPL
  SOURCE: s=userSource(id("graphdataset"))
  DATA: match_tot=col(source(s), name("match_tot"))
  DATA: inc_year=col(source(s), name("inc_year"))
  GUIDE: axis(dim(1), label("match_tot"))
  GUIDE: axis(dim(2), label("inc_year"))
  GUIDE: text.title(label("Simple Scatter with Fit Line of inc_year
by match_tot"))
```

```
  ELEMENT: point(position(match_tot*inc_year))
END GPL.
```

Stata-output 3.4

```
REGRESSION
  /DESCRIPTIVES MEAN STDDEV CORR SIG N
  /MISSING LISTWISE
  /STATISTICS COEFF OUTS R ANOVA
  /CRITERIA=PIN(.05) POUT(.10)
  /NOORIGIN
  /DEPENDENT inc_year
  /METHOD=ENTER match_tot.
```

Stata-output 3.5

```
REGRESSION
  /DESCRIPTIVES MEAN STDDEV CORR SIG N
  /MISSING LISTWISE
  /STATISTICS COEFF OUTS R ANOVA
  /CRITERIA=PIN(.05) POUT(.10)
  /NOORIGIN
  /DEPENDENT inc_year
  /METHOD=ENTER nation_dum.
```

SPSS has to the best of my knowledge no routine to show the labels of a categorical *x*-variable (or a dummy) in a regression similar to Stata's i.-routine.

Stata-output 3.6

When a categorical *x*-variable has more than two categories, such as the player position variable (pos) with four, we must first create in SPSS the number of categories we need and use them as a set of dummy variables. Here, we need three dummies: defender, midfielder, and attacker, making the goalkeepers the constant. The commands are:

```
RECODE pos (1=1) (ELSE=0) INTO defense.
RECODE pos (2=1) (ELSE=0) INTO midfield.
RECODE pos (3=1) (ELSE=0) INTO attack.
REGRESSION
  /MISSING LISTWISE
  /STATISTICS COEFF OUTS R ANOVA
  /CRITERIA=PIN(.05) POUT(.10)
  /NOORIGIN
  /DEPENDENT inc_year
  /METHOD=ENTER defense midfield attack.
```

Stata-output 3.7

```
REGRESSION
  /MISSING LISTWISE
```

```
/STATISTICS COEFF OUTS R ANOVA
/CRITERIA=PIN(.05) POUT(.10)
/NOORIGIN
/DEPENDENT infect_rate
/METHOD=ENTER price_sq_m.
```

Table 3.7

```
CROSSTABS
  /TABLES=health BY times_ex_gr
  /FORMAT=AVALUE TABLES
  /CELLS=COUNT COLUMN
  /COUNT ROUND CELL.
```

The command to get the ordinal correlation coefficient is:

```
CROSSTABS
  /TABLES=health BY times_ex_gr
  /FORMAT=AVALUE TABLES
  /STATISTICS=BTAU
  /CELLS=COUNT COLUMN
  /COUNT ROUND CELL.
```

Note that the value of 0.34 refers to Kendall's so-called Tau-b.

Table 3.8

```
CROSSTABS
  /TABLES=health BY gender
  /FORMAT=AVALUE TABLES
  /CELLS=COUNT COLUMN
  /COUNT ROUND CELL.
```

Figure 3.6

```
GGRAPH
  /GRAPHDATASET NAME="graphdataset" VARIABLES=alch_perc quality
MISSING=LISTWISE REPORTMISSING=NO
  /GRAPHSPEC SOURCE=INLINE
  /FITLINE TOTAL=YES.
BEGIN GPL
  SOURCE: s=userSource(id("graphdataset"))
  DATA: alch_perc=col(source(s), name("alch_perc"))
  DATA: quality=col(source(s), name("quality"))
  GUIDE: axis(dim(1), label("alch_perc"))
  GUIDE: axis(dim(2), label("quality"))
  GUIDE: text.title(label("Simple Scatter with Fit Line of quality
by alch_perc"))
  ELEMENT: point(position(alch_perc*quality))
END GPL.
```

Stata-output 3.8

```
REGRESSION
  /MISSING LISTWISE
  /STATISTICS COEFF OUTS R ANOVA
  /CRITERIA=PIN(.05) POUT(.10)
  /NOORIGIN
  /DEPENDENT youth_exe
  /METHOD=ENTER gender.
```

3.10 Chapter Exercises with Solutions

The exercises below use the data available for download on the book's website.

Exercises:

Exercise 1 (data: soccer, see appendix B of Chapter 2 for data documentation)
   1a   Describe the association between nation_dum and origin, if any.
   1b   Describe the association between nation_dum and match_tot, if any.
   1c   Describe the association between match_tot and age, if any.
   1d   Describe the association between pos and match_tot, if any.
   1e   Describe the association between club_rank and inc_year, if any.

Exercise 2 (data: student_exercise, see appendix C of Chapter 2 for data documentation)
   2a   Describe the association between fitness_cen and gender, if any.
   2b   Describe the association between times_exer and fitness_cen, if any.
   2c   Describe the association between econ and health_dum, if any.
   2d   Describe the association between hours_exer and times_exer, if any.
   2e   Describe the association between status and econ, if any.

Answers to Exercises (in Stata Only; see Section 3.9 for Equivalent SPSS Syntaxes):

Exercise 1 (data: soccer, see appendix B of Chapter 2 for data documentation)

   1a   Describe the association between nation_dum and origin, if any.

It makes most sense to treat origin as *x* and nation_dum as *y*. Furthermore, both variables are categorical/dummies. We thus make a cross-tabulation as in:

```
. tab nation_dum origin, col


           |       origin
nation_dum | Norwegian    foreign |     Total
-----------+----------------------+----------
        no |       127         33 |       160
           |     72.57      50.77 |     66.67
-----------+----------------------+----------
       yes |        48         32 |        80
           |     27.43      49.23 |     33.33
-----------+----------------------+----------
     Total |       175         65 |       240
           |    100.00     100.00 |    100.00
```

Among the Norwegian players, 27 percent have played one or more matches for the national team. The analogous percentage is 49 among the foreign players. There is a clear association between the two variables: Foreign players are national team players more often than Norwegian players.

1b   Describe the association between nation_dum and match_tot, if any.

It makes most sense to treat nation_dum as x and match_tot as y. Furthermore, x is categorical (a dummy) and y is continuous. We may thus do a one-way ANOVA or a regression; I start with the former:

```
. oneway match_tot nation_dum, t

                 |       Summary of match_tot
    nation_dum   |     Mean    Std. Dev.        Freq.
-----------------+-----------------------------------
             no  |   63.2875   54.939577          160
            yes  |  129.6375   98.093374           80
-----------------+-----------------------------------
          Total  |  85.404167  78.555787          240

                    Analysis of Variance
    Source              SS          df      MS             F     Prob > F
-----------------------------------------------------------------------------
Between groups      234790.533       1   234790.533      45.06    0.0000
 Within groups      1240081.26     238   5210.42547
-----------------------------------------------------------------------------
      Total         1474871.8      239   6171.0117

Bartlett's test for equal variances:  chi2(1) =   38.1579  Prob>chi2 = 0.000
```

The mean of total number of matches played in career is 85.40 for all the players; see the row for Total. The mean for the national team players is 130 matches, whereas the analogous mean is 63 for the players not having played for their national teams. Since 66 matches by all accounts is a marked difference (129.64 − 63.29 ≈ 66), we have a clear association between the two variables. The regression directly shows this 66-match difference along with the mean number of matches for players not having made appearances in their national teams (63.29), that is, the constant:

```
. reg match_tot i.nation_dum

      Source |       SS           df       MS      Number of obs   =       240
-------------+----------------------------------   F(1, 238)       =     45.06
       Model |  234790.533          1   234790.533  Prob > F        =    0.0000
    Residual |  1240081.26        238   5210.42547  R-squared       =    0.1592
-------------+----------------------------------   Adj R-squared   =    0.1557
       Total |  1474871.8         239   6171.0117  Root MSE        =    72.183

---------------------------------------------------------------------------------
   match_tot |      Coef.   Std. Err.      t    P>|t|     [95% Conf. Interval]
-------------+-------------------------------------------------------------------
  nation_dum |
         yes |      66.35   9.884102     6.71   0.000     46.8785      85.8215
       _cons |    63.2875   5.706589    11.09   0.000     52.04562     74.52938
---------------------------------------------------------------------------------
```

1c   Describe the association between match_tot and age, if any.

It only makes sense to treat age as x and match_tot as y. Furthermore, both variables are continuous. We thus do a regression as in:

```
. reg match_tot age

      Source |       SS           df       MS            Number of obs   =       240
-------------+----------------------------------         F(1, 238)       =    233.43
       Model |  730293.449          1   730293.449       Prob > F        =    0.0000
    Residual |  744578.347        238   3128.48045       R-squared       =    0.4952
-------------+----------------------------------         Adj R-squared   =    0.4930
       Total |    1474871.8        239   6171.0117       Root MSE        =    55.933


   match_tot |      Coef.   Std. Err.      t    P>|t|     [95% Conf. Interval]
-------------+----------------------------------------------------------------
         age |   12.05995   .7893392    15.28   0.000     10.50497    13.61494
       _cons |  -232.4761   21.11661   -11.01   0.000    -274.0755   -190.8768
```

The general interpretation of the regression coefficient is the average change in y given a one–unit increase in x. From this it follows that, say, a 25-year old player has played 12 more matches than a 24-year old player on average. There is a clear association between the two variables, as might be expected.

1d   Describe the association between pos and match_tot, if any.

It makes most sense to treat pos as x and match_tot as y. Furthermore, x is strictly categorical and y is continuous. We may thus do a one-way ANOVA or a regression; I start with the former:

```
. oneway match_tot pos, t

            |        Summary of match_tot
        pos |        Mean   Std. Dev.       Freq.
------------+------------------------------------
   goalkeepe |   61.565217   58.524918          23
    defense |   90.142857   71.381234          84
   midfield |   87.236842   85.553238          76
     attack |   85.596491   85.819965          57
------------+------------------------------------
      Total |   85.404167   78.555787         240

                        Analysis of Variance
    Source              SS          df      MS             F     Prob > F
------------------------------------------------------------------------
Between groups      15214.4018       3   5071.46727       0.82     0.4840
 Within groups      1459657.39     236   6184.98896
------------------------------------------------------------------------
      Total           1474871.8     239    6171.0117

Bartlett's test for equal variances:   chi2(3) =    6.6170   Prob>chi2 = 0.085
```

There is, save for the goalkeepers, very little variation in the number of matches played in the entire career: 90 (defense players), 87 (midfielders), and 86 (attackers).

The similar regression analysis contrasts the goalkeepers with the defense players, the midfielders, and the attackers. The analysis, of course, suggests there is no clear association between the two variables, save for a possible exception for the goalies standing out with less match experience.

```
. reg match_tot i.pos

      Source |       SS           df       MS      Number of obs   =       240
-------------+----------------------------------   F(3, 236)       =      0.82
       Model |   15214.4018         3   5071.46727   Prob > F       =    0.4840
    Residual |   1459657.39       236   6184.98896   R-squared      =    0.0103
-------------+----------------------------------   Adj R-squared  =   -0.0023
       Total |    1474871.8       239   6171.0117   Root MSE       =    78.645


   match_tot |      Coef.   Std. Err.      t    P>|t|     [95% Conf. Interval]
-------------+----------------------------------------------------------------
         pos |
     defense |   28.57764   18.50793     1.54   0.124    -7.884215    65.03949
    midfield |   25.67162   18.71614     1.37   0.171    -11.20043    62.54368
      attack |   24.03127   19.42733     1.24   0.217    -14.24186    62.30441
             |
       _cons |   61.56522   16.39855     3.75   0.000     29.25897    93.87146
```

1e   Describe the association between club_rank and inc_year, if any.

It only makes sense to treat club_rank as *x* and inc_year as *y*. Furthermore, inc_year is continuous and club_rank may be treated as continuous, although it is ordinal in a strict sense. We thus do a regression as in:

```
. reg inc_year club_rank

      Source |       SS           df       MS      Number of obs   =       240
-------------+----------------------------------   F(1, 238)       =     49.84
       Model |   1.9612e+11         1   1.9612e+11   Prob > F       =    0.0000
    Residual |   9.3660e+11       238   3.9353e+09   R-squared      =    0.1731
-------------+----------------------------------   Adj R-squared  =    0.1697
       Total |   1.1327e+12       239   4.7394e+09   Root MSE       =     62732


    inc_year |      Coef.   Std. Err.      t    P>|t|     [95% Conf. Interval]
-------------+----------------------------------------------------------------
   club_rank |  -6172.793    874.402    -7.06   0.000    -7895.349   -4450.237
       _cons |   136546.3   8180.645    16.69   0.000     120430.6      152662
```

A player representing the club finishing at, say, seventh place at the end of the season earns 6,200 Euros less on average than a player on the club finishing at sixth place. Better performing clubs thus pay better than clubs not performing so well, as might be expected. There is a marked association between the two variables.

Exercise 2 (data: student_exercise, see appendix C of Chapter 2 for data documentation)

2a   Describe the association between fitness_cen and gender, if any.

It only makes sense to treat gender as x and fitness_cen as y. Furthermore, both variables are categorical/dummies. We thus do a cross-table as in:

```
. tab fitness_cen gender, col

fitness_ce |        gender
        n |    female      male |     Total
-----------+----------------------+----------
        no |       210       108 |       318
           |     50.48     47.37 |     49.38
-----------+----------------------+----------
       yes |       206       120 |       326
           |     49.52     52.63 |     50.62
-----------+----------------------+----------
     Total |       416       228 |       644
           |    100.00    100.00 |    100.00
```

Almost 50 percent of the female students are members of fitness centers. The analogous percentage is almost 53 among the male students. There is thus no association between the two variables: Male and female students are fitness center members to the same extent for all practical purposes.

2b Describe the association between times_exer and fitness_cen, if any.

It arguably makes most sense to treat fitness_cen as x and times_exer as y, given the reasoning thus far in this book. Furthermore, x is a dummy and y is continuous. We may thus do a one-way ANOVA or a regression; I start with the former:

```
. oneway times_exer fitness_cen, t

             |       Summary of times_exer
fitness_cen |       Mean   Std. Dev.       Freq.
-----------+------------------------------------
        no |   2.3915094   1.5746484         318
       yes |   3.6687117   1.4741056         326
-----------+------------------------------------
     Total |   3.0380435   1.6520019         644

                      Analysis of Variance
    Source             SS        df      MS            F     Prob > F
-----------------------------------------------------------------------
Between groups       262.59       1    262.59      112.97     0.0000
Within groups     1492.22793     642   2.32434258
-----------------------------------------------------------------------
    Total         1754.81793     643   2.72911032

Bartlett's test for equal variances:  chi2(1) =   1.3948  Prob>chi2 = 0.238
```

The mean of exercise times per week is 3.04 among all students; see the row for Total. The mean for the fitness center members is 3.67 times, whereas the analogous mean is 2.39 for the non-members. Since 1.28 times per week is a large difference (3.67–2.39 = 1.28), we have a marked association between the two variables. The similar regression analysis directly shows this 1.28-times difference along with the mean of 2.39 for the non-members (i.e., the constant):

```
. reg times_exer i.fitness_cen

      Source |       SS           df       MS            Number of obs   =        644
-------------+----------------------------------         F(1, 642)       =     112.97
       Model |    262.59            1     262.59         Prob > F        =     0.0000
    Residual | 1492.22793          642  2.32434258        R-squared       =     0.1496
-------------+----------------------------------         Adj R-squared   =     0.1483
       Total | 1754.81793          643  2.72911032        Root MSE        =     1.5246


------------------------------------------------------------------------------
  times_exer |      Coef.   Std. Err.      t    P>|t|     [95% Conf. Interval]
-------------+----------------------------------------------------------------
 fitness_cen |
         yes |   1.277202    .120163    10.63   0.000     1.041242    1.513162
       _cons |   2.391509   .0854942    27.97   0.000     2.223627    2.559391
------------------------------------------------------------------------------
```

2c Describe the association between econ and health_dum, if any.

In the medical, social, and behavioral sciences you have no trouble finding evidence to suggest that financial trouble over a long period of time makes people less healthy. Conversely, if one becomes very sick, one has to stop working – leading most often to a worsened financial situation. That is, causation works both ways. I adopt the "material-istic" explanation and treat econ as x and health_dum as y. Furthermore, both variables are categorical/dummies. We thus do a cross-table as in:

```
. tab health_dum econ, col


           |                econ
health_dum |  not good         ok        good |     Total
-----------+---------------------------------+----------
    ok/good |        74        256         202 |       532
           |     90.24      86.20       76.23 |     82.61
-----------+---------------------------------+----------
  very good |         8         41          63 |       112
           |      9.76      13.80       23.77 |     17.39
-----------+---------------------------------+----------
      Total |        82        297         265 |       644
           |    100.00     100.00      100.00 |    100.00
```

Only 10 percent among those experiencing a not-so-good financial situation report very good health. Among those experiencing a good financial situation, the percent-age reporting good health is almost 24 – or more than twice as large. (The row for ok/good health of course shows the opposite pattern.) There is a clear association between the two variables: Compared with students experiencing a poorer financial situation, students experiencing a better such situation report having very good health more often.

2d Describe the association between hours_exer and times_exer, if any.

It only makes sense to treat `times_exer` as x and `hours_exer` as y. Furthermore, both variables are continuous. We thus do a regression as in:

```
. reg hours_exer times_exer

      Source |       SS           df       MS            Number of obs   =        644
-------------+----------------------------------         F(1, 642)       =    1895.95
       Model |  4759.14023          1   4759.14023        Prob > F        =     0.0000
    Residual |  1611.51932        642   2.5101547         R-squared       =     0.7470
-------------+----------------------------------         Adj R-squared   =     0.7466
       Total |  6370.65955        643   9.90771314        Root MSE        =     1.5843

------------------------------------------------------------------------------
  hours_exer |      Coef.   Std. Err.      t    P>|t|     [95% Conf. Interval]
-------------+----------------------------------------------------------------
  times_exer |   1.646828   .0378211    43.54   0.000     1.57256    1.721096
       _cons |  -.2228553    .130768    -1.70   0.089    -.4796399    .0339293
------------------------------------------------------------------------------
```

A student exercising for, say, three times per week exercises on average 1.65 more hours per week than a student exercising two times per week. There is a marked, positive association between the two variables, as is expected.

2e   Describe the association between `status` and `econ`, if any.

It makes most sense to treat status as x and econ as y. Furthermore, both variables are categorical. We thus make a cross-tabulation as in:

```
. tab econ status, col

           |              status
      econ |    single      bf/gf     cohabit |     Total
-----------+---------------------------------+----------
  not good |        43         24          15 |        82
           |     13.07      11.65       13.76 |     12.73
-----------+---------------------------------+----------
        ok |       153        101          43 |       297
           |     46.50      49.03       39.45 |     46.12
-----------+---------------------------------+----------
      good |       133         81          51 |       265
           |     40.43      39.32       46.79 |     41.15
-----------+---------------------------------+----------
     Total |       329        206         109 |       644
           |    100.00     100.00      100.00 |    100.00
```

As before, the frequency distribution for the financial situation variable among all students appears in the column on the right (`Total`). The main message is that the percentages referring to financial situation do not differ much among the three student statuses. There is no association between the two variables.

Notes

1 I emphasize again that we are assuming a causal direction; we are not claiming that x is in fact a cause of y. More on this in Sections 3.6, 3.7, and Chapter 4.

2 That said, I encourage you to spend more time on descriptive statistics to *really* get to know your data.

3 One-way refers to one *x*-variable, and ANOVA is short for ANalysis Of VAriance.

4 The other information in Stata-output 2.3 and SPSS-output 3.2 need not concern us at this point.

5 All correlations are associations, but not all associations are correlations. Hence, association is the more general term of the two.

6 Some might beg to differ claiming that we should do *correlation analysis* when studying correlations. Yet I see no reason to introduce correlation as a statistical technique per se, because regression does everything correlation does and much more.

7 I have more to say on regression analysis in Chapter 4, which draws heavily on my primer on regression analysis (Thrane, 2020). Consider this section a warm-up.

8 This equation should also include an error term, *e*, but we skip it for now. More on this in Chapters 4 and 6.

9 Regression assumes a causal direction from *x* to *y*, whereas correlation makes no such assumption. That is, the correlation could be $x \rightarrow y$ or $x \leftarrow y$ (reverse causation) or $x \leftarrow \rightarrow y$ (simultaneous causation). The commands in Stata and SPSS to produce *r* appear in Section 3.9.

10 A large amount of variation in many social science data sets often makes it hard to *see* the overall trend in an association between two variables; hence the need for a regression line to summarize this association.

11 In a bivariate regression, the constant is always the mean of *y* for *x* = zero if zero is a legitimate value in the data.

12 In addition to being a matter of personal taste, the choice between ANOVA and regression has much to do with different historical developments within different subjects. In psychology, pedagogy, and marketing, for example, ANOVA has a strong position. In economics, sociology, and political science, in contrast, regression is very dominant. Since I am a sociologist by training, I prefer regression. Sirkin (2005) has more on ANOVA.

13 Some might wonder how Stata or SPSS calculates b_0 and b_1. The calculus–procedure in question bears the name «Ordinary least squares» (OLS), but I see no reason to go into the statistical mechanics of this principle in the present context. Thrane (2020) guides you through the work if you choose to pursue it.

14 If the research question specifically concerns the differences between students above or below some theoretically interesting threshold age, say 22 years, we could similarly make a dummy.

15 Note that the ordered exercise frequency variable in Table 3.7 is a recoded version of the continuous exercise frequency variable in the data, that is, `times_exer`.

16 The value of six is a suggestion and it is not set in stone. Some would say seven or eight; others might say four or five. The demarcation value also depends on the variation in the *y* in question. For *y*-variables with little variation and/or much skewness, I personally move the threshold value upwards rather than downwards.

17 The second goal expressed less ambitiously: We want to find the *unique* effect of x_1 on *y*.

18 Provided that we are analyzing observational data. All the data in this book so far are observational. Matters may become easier when analyzing experimental data. I return to this in Section 3.7.

19 Similar reasoning leads to different effects for Carrie, Joni, Erica, and so on.

20 Another problem is that we cannot rule out that causation does not work the other way round, as in exercise hours (i.e., actual or planned) affecting the decision to buy a fitness center subscription. In principle, we could solve this by associating the membership dummy with exercise hours measured at some later point in time. The future cannot affect the past. This happens only in science-fiction movies.

21 The RCT is the gold standard among the quantitative research designs available for identifying causal effects in the social and behavioral sciences.

22 Variations of this quote appear throughout the Holmes stories, but it is most often attributed to *The Sign of the Four* (1890).

4 Associational Research Questions II

Multiple Regression

4.1 Introduction and Chapter Overview

This chapter introduces multiple regression analysis. Section 4.2 begins where Section 3.6 left off; we examine the association between x_1 and y while simultaneously controlling for a second and a third x-variable, x_2, and x_3. Section 4.3 then formalizes the classic multiple linear regression equation and discusses the linear regression *model* per se.

We have thus far been concerned with linear regression. That is, we have assumed that a straight regression line could summarize the association between x_1 and y. Sometimes, however, such a straight line is too simplistic a representation of how x_1 relates to y. More precisely, variable associations might be non-linear. Section 4.4 addresses how to handle non-linear variable associations within the multiple regression framework and explains when it is wise to do so.

Section 4.5 brings up another variant of such a too-simplistic-a-representation scenario: the case of parallel versus non-parallel regression lines for subgroups in the data. This section also brings up the question of when to prefer regression models with parallel regression lines and when to opt for non-parallel models.

We study experimental data using linear regression analysis in Section 4.6. More exactly, by replicating a classic psychology experiment, I show how statistical control typically is redundant when analyzing experimental data/RCT data.

Section 4.7 considers a dummy y-variable. In essence, this section swaps everything we have done in the book for a continuous y with that of a dummy y. The takeaway message is that we might consider adjusting the linear multiple regression model when y is a dummy, but that this oftentimes is not necessary. Section 4.8 summarizes the chapter and lists the key learning points, whereas Section 4.9 is the usual do-file and syntax-file coverage of commands. Section 4.10 provides some exercises with solutions.

Note! Throughout Chapter 4, my comments regarding the various associational results pertain only to what happens *within* the data. I do not refer to what might happen (or not) outside of the data. The latter inference topic is for Chapter 5.

4.2 Statistical Control for Observational Data: Two Examples

We have exclusively analyzed observational data in this book, save for a short bit on the workings of experimental data in Section 3.7. The vital feature of observational data is that the researcher is a passive observer of the data-generating process. For experimental data or RCT data, in contrast, the researcher actively manipulates this process. Survey data typically fall into the observational data category; the researcher handing out the

DOI: 10.4324/9781003252559-4

questionnaires has no bearing on the answers given.[1] We always face a vast challenge when analyzing observational data with the aim of finding the causal effect of x on y: to control for the effects of all other x-variables on y.

Section 3.6 introduced the need to control for other x-variables when trying to find the causal effect of fitness center subscription on hours of weekly exercise. Below we follow up on this by introducing a student data set similar to the one we have seen so far: student_exercise_motive. The data documentation appears in appendix A of this chapter. Three variables concern us at first: hours of weekly exercise (y), fitness center membership (x_1), and exercise motivation (x_2). These variables carry the names hours_exer, fitness_center, and exer_easy_motive. The descriptive statistics for the variables appear in Tables 4.1 and 4.2.

We have data on 510 students exercising for 3.92 hours per week on average; cf. Table 4.1. The standard deviation is 3.70, and the range is 23 hours ($23 - 0 = 23$). Panel A in Table 4.2 shows that 65 percent of the students presently are fitness center members, making the non-member category 35 percent. The exercise motivation variable is ordinal, that is, a statement shedding light on the students' varying exercise motivation. The statement reads, 'I easily find the motivation to exercise!' (totally disagree $= 1$, disagree $= 2$, neither disagree nor agree $= 3$, agree $= 4$, and totally agree $= 5$). Panel B in Table 4.2 shows that most students are on the agreeing side of the statement. Yet as much as 30 percent of the students are still on the disagreeing side ($12 + 18 = 30$).[2]

We first associate the center membership dummy to the exercise variable in a bivariate regression. Stata-output 4.1 shows the results. The analogous SPSS-output tells the same story and is thus redundant. This is also the case for most of the Stata-outputs in

Table 4.1 Descriptive statistics for hours of weekly exercise in the student exercise motive data.

Variable:	$N =$	Mean	SD	Min.	Max.
Hours of weekly exercise	510	3.92	3.70	0	23

Table 4.2 Descriptive statistics for fitness center membership (Panel A) and exercise motivation (Panel B) in the student exercise motive data. $N = 510$.

Variables:	Frequency	Percent
Panel A		
Fitness center member:		
No	180	35
Yes	330	65
Panel B:		
Exercise motivation:[a]		
Totally disagree (1)	62	12
Disagree (2)	91	18
Neither/nor (3)	140	27
Agree (4)	136	27
Totally agree (5)	81	16

[a] Exercise motivation statement: 'I easily find the motivation to exercise!'

this chapter. In those few instances where SPSS differs from Stata, however, I will also present the analogous SPSS-outputs.

```
. reg hours_exer i.fitness_center

      Source |       SS           df       MS          Number of obs   =       510
-------------+----------------------------------       F(1, 508)       =     50.02
       Model |  625.227458         1  625.227458       Prob > F        =    0.0000
    Residual |  6349.14558        508  12.4983181       R-squared       =    0.0896
-------------+----------------------------------       Adj R-squared   =    0.0879
       Total |  6974.37304        509  13.7021081       Root MSE        =    3.5353

-----------------------------------------------------------------------------------
 hours_exer |      Coef.   Std. Err.       t    P>|t|     [95% Conf. Interval]
-------------+---------------------------------------------------------------------
fitness_center |
        yes |   2.316919   .3275802     7.07   0.000     1.67334    2.960498
      _cons |   2.419444   .2635054     9.18   0.000     1.90175    2.937139
-----------------------------------------------------------------------------------
```

Stata-output 4.1 Results for regression of hours of exercise by fitness center membership in the student exercise motive data.

Fitness center members exercise on average 2.32 more hours per week than non–members, which coincidentally is close to our imagined result in Section 3.6. The non–members exercise 2.42 hours on average, as suggested by the constant. Yet, since we expect that the membership group includes many students with excess exercise motivation and, conversely, that the non–member group includes many students lacking exercise motivation, we suspect that the membership coefficient by no means is causal. We thus want to compare members and non–members of fitness centers having *similar* levels of exercise motivation with respect to exercise hours per week. This like–for–like comparison, to simplify it somewhat, is what multiple regression does. Stata-output 4.2 presents the multiple regression in question.

```
. reg hours_exer i.fitness_center exer_easy_motive

      Source |       SS           df       MS          Number of obs   =       510
-------------+----------------------------------       F(2, 507)       =     89.47
       Model |   1819.4585         2  909.729252       Prob > F        =    0.0000
    Residual |  5154.91453        507  10.1674843       R-squared       =    0.2609
-------------+----------------------------------       Adj R-squared   =    0.2580
       Total |  6974.37304        509  13.7021081       Root MSE        =    3.1886

-----------------------------------------------------------------------------------
 hours_exer |      Coef.   Std. Err.       t    P>|t|     [95% Conf. Interval]
-------------+---------------------------------------------------------------------
fitness_center |
        yes |   1.327498    .309243     4.29   0.000    .7199428    1.935054
exer_easy_motive |   1.290549   .1190795    10.84   0.000    1.056599    1.524499
      _cons |   -1.02202   .3966372    -2.58   0.010   -1.801275   -.2427647
-----------------------------------------------------------------------------------
```

Stata-output 4.2 Results for multiple regression of hours of exercise by fitness center membership and exercise motivation in the student exercise motive data.

The multiple regression coefficient for the fitness center dummy is 1.33 hours. Comparing students with similar levels of exercise motivation, we find that members of a fitness centers on average exercise 1.33 more hours per week than non–members. We are now certain that some part of the bivariate effect of a fitness center membership (i.e., 2.32) was not causal. We may even claim that about 43 percent of this bivariate effect was brought about by exercise motivation $((2.32 − 1.33)/2.32 = 0.427)$. Yet, since we still find a difference of more than one hour of exercising per week between members and non-members who are similar regarding exercise motivation, exercise motivation does not seem to account for the complete bivariate group difference.

The multiple regression in Stata-output 4.2 finds the effect of the fitness center dummy on exercise hours while simultaneously taking the exercise motivation variable into account. Complicated math takes care of this. Metaphorically, we may think of the statistics program first doing a regression between the membership dummy and exercise hours for the students *totally disagreeing* with the motive statement. Next, the statistics program does a second and similar regression for the students *disagreeing* with the motive statement, and so on three more times for the remaining three levels of motivation: neither/nor, agree, and totally agree. Finally, the program finds the average of these five dummy coefficients, which becomes 1.33 in our case. This keeping-another-*x*-variable-fixed ability is the pillar of multiple regression. It is also the foundation of the holding constant clause; the expression that members of fitness centers on average exercise 1.33 more hours per week than non-members *holding* exercise motivation *constant*. As an alternative expression, we might say that members of fitness centers on average exercise 1.33 more hours per week than non-members *controlling for* exercise motivation.

The multiple regression coefficients for exercise motivation is 1.29. This suggests that students answering, say, *agree* (4) on the statement exercise 1.29 more hours per week on average than students answering *neither/nor* (3) holding the fitness center membership dummy constant;[3] that is, in multiple regression each *x*-variable is simultaneously controlled for every other *x*-variable included in the analysis.[4]

Having explained the principle of holding constant or controlling for with regard to the three-variable case, it is straightforward to imagine the same principle in the four, five, and six-variable case, and so on. Let us put this into practice. Our data also contain information paralleling the motive variable for the statement, 'Exercising three times per week is important for my quality of life!' What happens if we add this importance-variable, x_3, to the multiple regression? Stata-output 4.3 shows the results.

```
. reg hours_exer i.fitness_center exer_easy_motive exer_imp_qol

      Source |       SS           df       MS        Number of obs   =       510
-------------+----------------------------------    F(3, 506)       =     74.72
       Model |  2141.1789          3   713.726298    Prob > F        =    0.0000
    Residual |  4833.19414       506   9.55176708    R-squared       =    0.3070
-------------+----------------------------------    Adj R-squared   =    0.3029
       Total |  6974.37304       509   13.7021081    Root MSE        =    3.0906

------------------------------------------------------------------------------------
    hours_exer |      Coef.   Std. Err.      t    P>|t|     [95% Conf. Interval]
-----------------+------------------------------------------------------------------
 fitness_center |
            yes |   .825015   .3119878     2.64   0.008     .2120641    1.437966
exer_easy_motive |  .8575133   .1374359     6.24   0.000     .587498     1.127529
    exer_imp_qol |  .7548036   .1300578     5.80   0.000     .4992837    1.010323
           _cons | -2.127475   .4290406    -4.96   0.000    -2.970395   -1.284555
------------------------------------------------------------------------------------
```

Stata-output 4.3 Results for multiple regression of hours of exercise by fitness center membership, exercise motivation, and exercise importance in the student exercise motive data.

Stata-output 4.3 suggests that fitness center members on average exercise 0.83 more hours per week than non-members when holding constant, or controlling for, exercise motivation *and* exercise importance. When comparing fitness center members and non-members equally in terms of exercise motivation and exercise importance, we find that the former group exercises just short of one hour more per week than the latter group. Note also the reduction in the effect of exercise motivation from 1.29 to 0.86 once controlled for the importance variable. The importance variable has about the same effect on hours of exercise as the motivation variable: 0.75 versus 0.86. Motivation and importance in combination account for 64 percent of the bivariate association between the fitness center dummy and hours of exercise ((2.32–0.83)/2.32 = 0.642). Yet exercise motivation and exercise importance are not able to rule out completely that a fitness center membership by itself has a larger-than-zero causal effect on exercise hours. That is, we find an almost one-hour difference in exercise hours per week when comparing members and non-members who are equal in terms of exercise motivation and exercise importance.

I could go on and on entering more x-variables like this in the regression. Yet I quit now since nothing essentially new would happen, save for the fact that the regression coefficient for the fitness center dummy might decrease or increase in size when a fourth x-variable enters the regression, and so on for a fifth and a sixth x-variable.[5] The upshot is that owing to the statistical control ability, multiple regression finds the unique effect of each x-variable in terms of explaining variation in the y-variable. The multiple regression coefficients expresses this adjusted effect. That said, multiple regression only controls for the x-variables included in the regression. We are, therefore, certain about the size of the regression coefficient for x_1, x_2, and so on only to the extent that the regression includes all x-variables responsible for bringing about variation in y. This makes the statistical control approach less effective and less certain than the experimental control method in terms of finding causal effects.

Some learn lots from numbers in tables; others learn better with graphs. In the paragraphs to come, we take on the same problem as above from a visual perspective. We have analyzed the Christmas beer data on several occasions by now; Figure 4.1 presents the correlation between taste quality (x_1) and bottle price (y) along with the regression line. Stata-output 4.4 presents the results yielding the plot in Figure 4.1.

```
. reg price quality

      Source |       SS           df       MS          Number of obs   =        75
-------------+----------------------------------        F(1, 73)        =      8.02
       Model |  13.3826297         1   13.3826297       Prob > F        =    0.0060
    Residual |  121.81737         73   1.6687311        R-squared       =    0.0990
-------------+----------------------------------        Adj R-squared   =    0.0866
       Total |       135.2        74   1.82702703       Root MSE        =    1.2918


-------------------------------------------------------------------------------
       price |      Coef.   Std. Err.      t    P>|t|     [95% Conf. Interval]
-------------+-----------------------------------------------------------------
     quality |   .1488282   .0525542     2.83   0.006     .0440878    .2535686
       _cons |   5.474343   .3212263    17.04   0.000      4.83414    6.114546
-------------------------------------------------------------------------------
```

Stata-output 4.4 Results producing Figure 4.1.

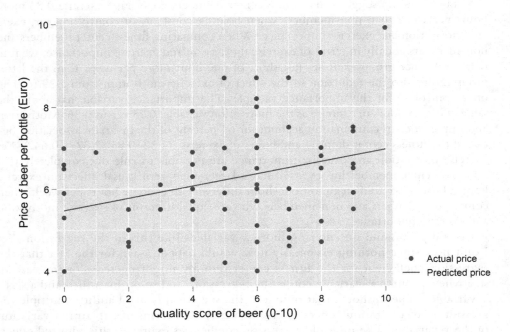

Figure 4.1 Scatterplot of correlation between quality score of beers and bottle price of beers, with regression line.

It seems as though we get what we pay for in terms of taste quality when paying more for a beer. That is, we find a positive correlation between the variables. The taste coefficient suggests that a beer scoring a six on taste costs almost 0.15 Euros more than a beer obtaining a score of five on average. Yet this quality coefficient suffers from the same problem as the fitness center dummy variable we just saw; the size of 0.15 might say little or nothing about taste quality's causal effect on price. To shed more light on this, we include the alcohol level variable (x_2) in the analysis and run a multiple regression. Stata-output 4.5 displays the results.

```
. reg price quality alch_perc

      Source |       SS           df       MS            Number of obs   =        75
-------------+----------------------------------         F(2, 72)        =     21.89
       Model | 51.1239659          2   25.561983         Prob > F        =    0.0000
    Residual | 84.0760341         72   1.1677227         R-squared       =    0.3781
-------------+----------------------------------         Adj R-squared   =    0.3609
       Total |      135.2         74  1.82702703         Root MSE        =    1.0806

-------------------------------------------------------------------------------------
       price |      Coef.   Std. Err.       t    P>|t|     [95% Conf. Interval]
-------------+-----------------------------------------------------------------------
     quality |  -.0172467   .0527833     -0.33   0.745    -.1224683    .0879748
   alch_perc |   .5476689   .0963339      5.69   0.000     .3556307    .739707
       _cons |    1.88978   .6853899      2.76   0.007      .52348    3.25608
-------------------------------------------------------------------------------------
```

Stata-output 4.5 Results for regression of beer price by quality and alcohol level in the Christmas beer data.

The quality coefficient is close to zero, and even negative, in the multiple regression. Figure 4.2 shows the difference in the quality effect for the bivariate and multiple regression. What is happening here? The answer is that the bivariate regression between quality and price does not take other x-variables, such as alcohol level, into account. In contrast, the multiple regression examines the quality-price association for beers having the same alcohol level.[6] This multiple regression suggests no linear association between quality and price. That is, we do not get what we pay for in terms of better taste when paying more for a beer! Why this new and contradictory conclusion? The following scenario explains the major change in the taste quality effect: We saw in Figure 3.6 a positive correlation between alcohol level and taste quality: Higher alcohol levels implied better taste scores. Stata–output 4.5 and Figure 3.3 showed a positive correlation between alcohol level and bottle price: Higher alcohol levels suggested pricier beers. Combined, these two positive correlations suggest the causal setup in Figure 4.3: The alcohol level variable is the common cause, or confounder, of the *spurious* (false) relationship between taste quality and price shown in Figure 4.1. The relationship between taste quality and price is 100 percent spurious or false in this example. In the fitness center membership dummy and exercise hours example, in contrast, the relationship was only partially spurious/false.

Having explained the metaphorical mechanics of multiple regression, we are now ready to address some new but related topics.

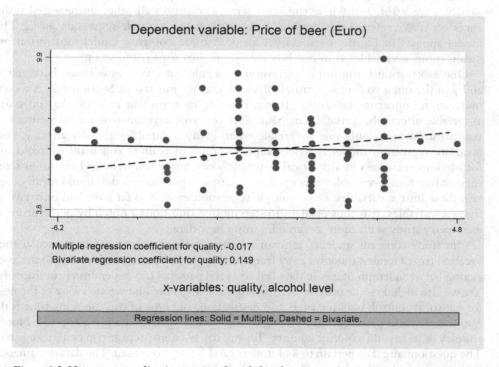

Figure 4.2 How taste quality is associated with bottle price in bivariate regression (dashed line) and *not* associated with bottle price in multiple regression (solid line).

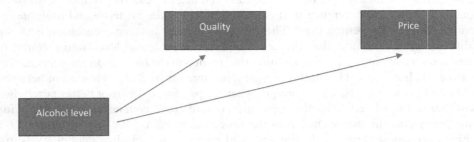

Figure 4.3 Alcohol level is associated with taste quality and bottle price (i.e., two direct arrows), but there is no association (i.e., no direct arrow) between taste quality and bottle price. The alcohol level variable is the confounder or common cause behind the spurious (false) taste quality-bottle price relationship.

4.3 The Multiple Regression Model and R^2

I have kept an informal approach to regression and all things statistical so far. I intend to keep up this easy-going approach, but some formalities are still necessary. First up, we write the multiple regression equation properly as

$$y = b_0 + b_1 x_1 + b_2 x_2 + \dots b_n x_n + e,$$

where subscript n implies that we may have as many regression coefficients and x-variables we want, and where the error term, e, captures all other unmeasured influences on y. We refer to this regression equation as the multiple regression *model*. This model appears in Figure 4.4 in visual form. A more complex model most often just means more x-variables as in one box for x_3, one box for x_4, and so on.

One motivation for multiple regression is to rule out a false association between x_1 and y in the quest to find x_1's causal effect on y, as we just saw in Section 4.2. A second motivation concerns plausibility. Is it reasonable to expect in real life that only one x-variable affects the variation in your y? If yes, you have no worries. In contrast, if you insist on using only one x-variable when many x-variables actually affect y, your bivariate regression model is a poor representation of reality. You want to avoid this for obvious reasons. On a practical note: If theory, prior research, and common sense suggest that four x-variables affect y, your multiple regression model should ideally contain these four x-variables. Our multiple regression models so far have had only two or three x-variables. It is thus prudent to step up our ambitions a notch by performing an exemplary study with more x-variables using new data.

Our study concerns students' tourism activities during summer and in particular their vacation trip of longest duration away from home. Our research question refers to the association between length of stay in days (x_1) and total personal trip expenditures in Euros (y). Yet we also include some other x-variables in the analysis because we are aware of the need for statistical control: booking time, trip destination, and type of trip. Booking time is the number of weeks passing from booking to trip start. The destination is a trip to a Nordic country or to beyond a Nordic country. Type of trip is a non-package trip or a package trip. The questionnaire data pertain to 444 students and (stud_tourism). The data documentation is at the end of the chapter (Appendix B). Table 4.3 is a typical way of presenting summary descriptive statistics for the five variables in question in a thesis or a research paper.

The information for the two key variables appears on the top two rows of the table. We note that the average trip incurs almost 600 Euros in total expenditures and lasts for

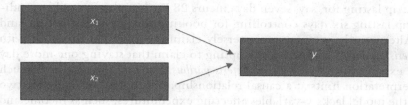

Figure 4.4 Multiple regression model containing two independent variables, x_1 and x_2.

Table 4.3 Descriptive statistics for tourism study variables.

Variables:	N =	Mean	SD	Min.	Max.
Total trip expenditures (Euros)	444	595.1	597.2	0	2,500
Length of stay (in days)	444	10.1	8.5	1	60
Booking time (in weeks)	444	7.9	9.0	0	50
Destination (1 = Beyond Nordic country)	444	0.62	0.49	0	1
Type of trip (1 = Package trip)	444	0.35	0.48	0	1

about ten days. The mean of booking time is eight weeks. The rows for the dummies deserve special attention since I previously claimed that the mean does not make sense for categorical variables. This was only partially correct because the mean of a dummy is the *proportion* of units with the value of one. That is, 62 percent of the students went on trips to destinations beyond the Nordic countries. Similarly, 35 percent of the students took part in package trips.

Formally, the multiple linear regression model is

$$y = b_0 + b_1x_1 + b_2x_2 + b_3x_3 + b_4x_4,$$

where y is total trip expenditures in Euros (tot_spend), x_1 is length of stay in days (los), x_2 is booking time in weeks (book_time), x_3 is trip destination (destin), and x_4 is type of trip (type_trip). I omit the error term. Stata–output 4.6 presents the results of the regression model.

```
. reg tot_spend los book_time i.destin i.type_trip

      Source |       SS           df       MS            Number of obs   =       444
-------------+----------------------------------         F(4, 439)       =     61.66
       Model |  56832535.8          4    14208134        Prob > F        =    0.0000
    Residual |   101160825        439  230434.681        R-squared       =    0.3597
-------------+----------------------------------         Adj R-squared   =    0.3539
       Total |   157993361        443  356644.156        Root MSE        =    480.04

------------------------------------------------------------------------------
   tot_spend |      Coef.   Std. Err.      t    P>|t|     [95% Conf. Interval]
-------------+----------------------------------------------------------------
         los |   27.60737    2.79562     9.88   0.000     22.11291    33.10184
   book_time |   5.941777   2.701041     2.20   0.028     .6331987    11.25035
             |
      destin |
Beyond Nordic |   445.487   58.94681     7.56   0.000     329.6339      561.34
             |
   type_trip |
Package trip |  -74.39578   56.88282    -1.31   0.192    -186.1923     37.4007
       _cons |   20.01848   44.93898     0.45   0.656    -68.3038     108.3408
------------------------------------------------------------------------------
```

Stata-output 4.6 Total trip expenditures by independent variables for the student tourism data.

We note that the coefficient of main interest, length of stay or los, is almost 28: On average, a trip lasting for, say, seven days incurs 28 Euros more in total expenditures than a trip lasting six days controlling for booking time, trip destination, and type of trip. Alternatively, we may more tersely claim the same but finish off with '... six days *ceteris paribus*.'[7] It is perhaps tempting to claim that staying one more day increases total expenditures by 28 Euros *ceteris paribus*, but I advise against it. Such a dynamic interpretation hints at a causal relationship, but this is not likely for two reasons: First, the model lacks x-variables affecting expenditures, such as income and savings. Second, length of stay is measured at the same time as expenditures. It is, therefore, possible that disposable total expenditures could curb length of stay as in reverse causation.

Trips booked, say, ten weeks in advance incur on average six Euros more in total expenditures than trips booked nine weeks in advance *ceteris paribus*. A 20-week difference thus amounts to 120 Euros. Compared with a trip to the Nordic countries, a trip to the non–Nordic countries on average entails 446 Euros more in total expenditures *ceteris paribus*. In contrast, package trips incur on average only 74 Euros less than non–package trips *ceteris paribus*.

That was it for the individual x-variables and their individual *ceteris paribus* contributions to the statistical explanation of total trip expenditures. What about the regression model per se? To what extent do the four x-variables in combination explain variation in total trip expenditures? R^2 (R-squared) answers this question.

4.3.1 R-squared (R^2)

In regression, we are often concerned about how "good" our regression model is. We might also in this regard speak of a good model fit. Formally, this boils down to how much of the variation in y the x-variables explain combined. A poor regression model (a model with poor fit) explains little or nothing of the variation in y; a good or better-fitting model explains more and preferably much more. R^2 formalizes this assessment, yielding a number between zero and one. R^2 for the model in Table 4.6 is 0.36 or 36 percent; note the R-squared in the upper right of the output. The four x-variables in the regression model combined explain 36 percent of the variation in total trip expenditures.[8] The error term, e, and randomness account for the remaining 64 percent.

The intuition behind R^2 is straightforward. There is variation in y irrespective of the x-variables. We refer to this total variation in y as the total sum of squares. For the regression model in Table 4.6, this Total amount is 157993361; cf. the upper left of the output (SS is short for sum of squares). Similarly, we have the variation in y accounted for by the x-variables in the regression model. For the regression model in Table 4.6, this Model amount is 56832535.8; cf. two rows above Total SS. Dividing 56832535.8 by 157993361, we get 0.36: the variation in y accounted for by the four x-variables.

We find R^2 at the top of SPSS-output 4.1. Similarly, we find the total and model sum of squares in the middle part of the output. Regarding the effects of the dummy variables destin and type_trip, we note that SPSS does not present the categories coded as one for the dummies, that is, beyond Nordic trip and Package trip. Otherwise, the results are similar except for rounding.

There is another way to look at R^2. Imagine guessing on total trip expenditures for a random student. In the long run, as the statisticians are fond of saying, your best guess would be the average trip expenditures. If you base your guess on the regression model, however, it will be 36 percent better in the long run than guessing on the average.[9]

Model Summary

Model	R	R Square	Adjusted R Square	Std. Error of the Estimate
1	,600[a]	,360	,354	480,036

a. Predictors: (Constant), type_trip, book_time, los, destin

ANOVA[a]

Model		Sum of Squares	df	Mean Square	F	Sig.
1	Regression	56832535,84	4	14208133,96	61,658	,000[b]
	Residual	101160825,1	439	230434,681		
	Total	157993361,0	443			

a. Dependent Variable: tot_spend

b. Predictors: (Constant), type_trip, book_time, los, destin

Coefficients[a]

Model		Unstandardized Coefficients		Standardized Coefficients		
		B	Std. Error	Beta	t	Sig.
1	(Constant)	20,018	44,939		,445	,656
	los	27,607	2,796	,391	9,875	,000
	book_time	5,942	2,701	,090	2,200	,028
	destin	445,487	58,947	,363	7,557	,000
	type_trip	-74,396	56,883	-,060	-1,308	,192

a. Dependent Variable: tot_spend

SPSS-output 4.1 Total trip expenditures by independent variables for the student tourism data.

Many students want to know if their regression model "is good," "has good fit," or has a "satisfactory R^2." There is no clear-cut answer to such questions. The best one is probably that it depends. That is, the R^2 of a particular regression model should be judged against a yardstick we oftentimes find in prior research. No more, no less. The importance of R^2 has also much to do with the nature of the research question. R^2 is often uninteresting if the primary interest lies in x_1's causal effect on y. In contrast, R^2 is more relevant if we want to know how five x-variables in combination affect y. That said, the idea of 100 percent explained variance is no benchmark in most cases. Moreover, an R^2 of, for example, 15 percent may be satisfactory for regression models based on survey questionnaire data.

4.4 Non-Linear Effects

Straight regression lines have an aesthetic and orderly appeal, and the same goes for straight, parallel regression lines. Yet the world around us is often messier than this. To account for the complexities of real life, we must often adjust our regression models to become more attuned to how reality works and sometimes bites. This section and the next, that is, 4.5, is about doing so.

4.4.1 Non–Linear Effects 1: Quadratic Regression

We have performed only linear regression in this book so far; all our regression lines for continuous x-variables have been straight. This simplification is useful for many purposes. Yet sometimes it is a too simplistic way of associating x_1 and y. Think of the association between age and income among athletes, and suppose a researcher has found a positive correlation for these variables in a group of hockey players. The usual interpretation of this correlation would be the more of the former (age), the more of the latter (income). But does this make sense in real life? Is it reasonable that as players get older and older they get to earn more and more money by each passing year? Not so much. Let us dive into this scenario for our soccer players.

The scatterplot and regression line for the age of the soccer players and their yearly income appear in Figure 4.5. We note the positive correlation, that is, an upward–sloping regression line in sync with the fictitious hockey researcher's results. The regression coefficient for the age variable is 2,502 Euros (not shown) and suggests that a 25–year old player on average earns 2,502 Euros more than a 24–year old player. Again, we should

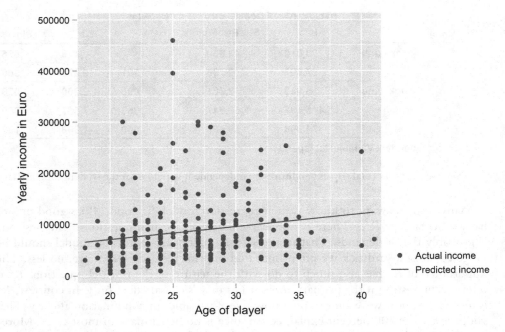

Figure 4.5 Scatterplot of correlation between age of player and yearly income, with linear regression line.

not say, "Aging by one year implies 2,502 Euros more in yearly income on average." This is too close to a causal interpretation that cannot be justified here.

Two concerns are important regarding Figure 4.5. One is theoretical and relates to prior research; the other concerns the empirical plot itself. Theoretically and in accordance with prior studies, we should expect an inverse U-association between age and income: Income tends to rise steeply at the start of the career, flatten out in midlife, and then decrease in later years before retirement. Given that an athlete's career is a micro-version of this life cycle, a linear regression line is dubious. Empirically, the plot shows that many players are located far away from the regression line. Moreover, many of the players receiving the largest incomes appear in the middle of the plot – in step with the theoretically expected inverse U-pattern. The long and short of this is that the linear regression does not sit well with either theory or data. An R^2 of 2.78 percent (not shown) also reflects this poor model fit. These shortcomings beg the question of how to make a regression line more in accordance with the inverse U-pattern. The answer is a quadratic regression model, and this model appears in Figure 4.6.

The quadratic regression line has the inverse U-pattern in sync with the theoretical prediction. It also has a better model fit than the linear model; R^2 for the quadratic model is 5.3 percent; cf. Stata–output 4.7. According to the *non-linear* regression having the inverse U-pattern, the soccer players' incomes peak at around 30 years of age. Now for some technicalities in this regard.

The quadratic or curved regression line in Figure 4.6 is the result of a regression with two x-variables: age (i.e., age in the data) and age^2 (i.e., age-square, which is not in the data). That is, to estimate the regression model bringing about Figure 4.6, we must first create the variable age^2 and then add this new variable to the regression responsible for

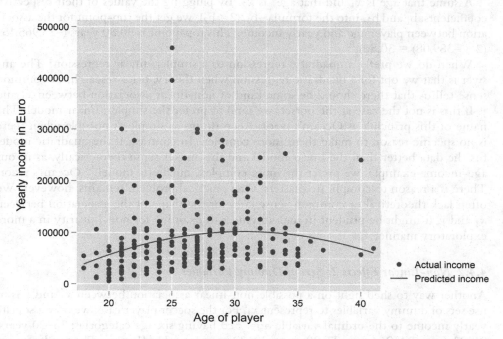

Figure 4.6 Scatterplot of correlation between age of player and yearly income, with quadratic regression line.

Figure 4.5. Fortunately, Stata has a built-in command that does this in one step. (Section 4.9 shows the similar procedure in SPSS.) The output appears in Stata-output 4.7.

```
. reg inc_year c.age##c.age

      Source |       SS           df       MS            Number of obs   =       240
-------------+------------------------------            F(2, 237)       =      6.27
       Model |   5.6961e+10          2   2.8480e+10      Prob > F        =    0.0022
    Residual |   1.0758e+12        237   4.5390e+09      R-squared       =    0.0503
-------------+------------------------------            Adj R-squared   =    0.0423
       Total |   1.1327e+12        239   4.7394e+09      Root MSE        =     67372


--------------------------------------------------------------------------------------
    inc_year |      Coef.   Std. Err.      t    P>|t|     [95% Conf. Interval]
-------------+------------------------------------------------------------------------
         age |   23965.15   9102.132     2.63   0.009     6033.732     41896.57
             |
c.age#c.age |  -389.0802    164.103    -2.37   0.019    -712.3671    -65.79337
             |
       _cons |  -266856.6   123799.9    -2.16   0.032    -510745.3    -22967.83
--------------------------------------------------------------------------------------
```

Stata-output 4.7 Results producing Figure 4.6.

The regression coefficients for age and c.age#c.age, which is Stata's technical term for age^2, are 23,965.15 and −389.08. In most applications, however, the main concern is just their signs. Because the age^2 coefficient has a negative sign, we know *a priori* that the regression line has an inverse U-shape.

Assume that age is x_1 and that age^2 is x_2. By plugging the values of their respective coefficients, b_1 and b_2, into the formula $-b_1/(2 \times b_2)$, we get the top-point for the association between player age and yearly income. This top-point is 30.80 years ($-23,965.15/ (2 \times -389.08) = 30.80$).[10]

When do we prefer a quadratic regression to a simpler, linear regression? The answer is that we opt for a quadratic regression when theory, prior research, or common sense tell us that there should be some kind of non-linear association between x_1 and y. If this is not the case at the outset, we tend to prefer the simpler, linear model. The name of this principle is Occam's razor: Keep things as simple as possible when there is no specific reason to make them more complex. In contrast, if the quadratic model fits the data better than the linear model and this makes sense theoretically, as in our age-income example, we prefer the more complex, quadratic model.[11] Occam's razor: There *is* a reason to complicate matters. One practical problem remains, however: We often lack theoretical or common-sense cues for the shape of the association between x_1 and y. It might be prudent in such instances to examine for non-linearity in a more exploratory manner.

4.4.2 Non-Linear Effects 2: Sets of Dummy Variables

Another way to shed light on a possible non-linear association between x_1 and y is to use sets of dummy variables to represent x_1. For the soccer player case, we may associate yearly income to the ordinal variable age_ord having six age categories: 18–20 years, 21–23 years, 24–26 years, 27–29 years, 30–32 years, and 33–41 years. The results appear in Stata-output 4.8.

```
. reg inc_year i.age_ord

      Source |       SS           df       MS            Number of obs   =      240
-------------+------------------------------            F(5, 234)       =     3.35
       Model |  7.5617e+10         5  1.5123e+10        Prob > F        =   0.0061
    Residual |  1.0571e+12       234  4.5175e+09        R-squared       =   0.0668
-------------+------------------------------            Adj R-squared   =   0.0468
       Total |  1.1327e+12       239  4.7394e+09        Root MSE        =    67212


------------------------------------------------------------------------------
    inc_year |      Coef.   Std. Err.      t    P>|t|     [95% Conf. Interval]
-------------+----------------------------------------------------------------
     age_ord |
       21-23 |   37513.53   17585.07     2.13   0.034     2868.233    72158.82
       24-26 |    55798.8   16970.06     3.29   0.001     22365.17    89232.43
       27-29 |   57758.29   17378.01     3.32   0.001     23520.95    91995.63
       30-32 |   68161.29    18875.7     3.61   0.000     30973.25    105349.3
       33-41 |   44939.06   19895.24     2.26   0.025     5742.388    84135.74
             |
       _cons |    38165.6   14666.96     2.60   0.010     9269.433    67061.78
------------------------------------------------------------------------------
```

Stata-output 4.8 Results for regression of yearly income by an ordered age variable in the soccer player data.

The age group not showing up in the results window is the reference: the 18–20 years group. This youngest age group on average earns 38,166 Euros yearly, that is, the constant. Compared with this, the 21–23 years group earns 37,514 Euros more on average. The same interpretation applies to the remaining age groups, and we note at first the steady age increase in income. The non-linearity of the association is apparent for the two oldest groups: Whereas the 30–32 years old players earn 68,161 Euros more than the youngest players on average, the analogous difference between the oldest and the youngest players is only 44,393. We note the same overall trend as in Figure 4.6.

4.4.3 Non-Linear Effects 3: Logarithmic Regression

The two prior procedures for handling non-linearity have one thing in common: They adapt the regression to a non-linear data pattern. The next procedure, in contrast, first changes the data by linearizing a non-linear pattern and then runs a linear regression in the usual manner. I use a new data set on red wines to illustrate (red_wine); cf. appendix C of this chapter for documentation. The y is bottle price in Euros (mean = 32.1, SD = 35.3, min = 7, and max = 252; not shown) and the x_1 is taste quality. The taste quality variable is the Parker scale, with a possible range from 60 points (terrible) to 100 points (perfect). The mean is 84.4 and the SD is 5.9 (not shown). The data comprise 218 wines. We expect a positive correlation between taste quality and bottle price based on prior studies. Figure 4.7 presents both the linear and the quadratic regression model for the two variables. The main message of the figure is that neither regression model line fits the data very well, although the quadratic model possibly looks slightly better in terms of model fit.[12]

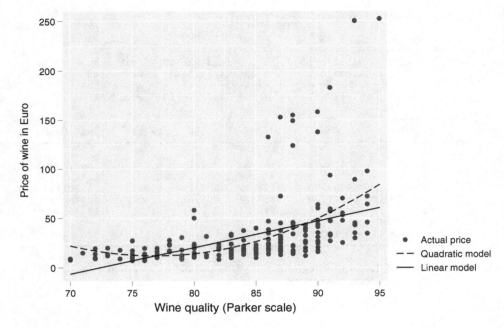

Figure 4.7 Scatterplot of correlation between taste quality and price of red wine, with regression lines based on linear (solid line) and quadratic (dashed line) model specifications.

The quadratic regression model yields by its inherent mathematical properties a regression line with either a top-point or a low-point. When the data pattern has no such point, however, it follows that the quadratic regression is not optimal. We, therefore, want a better solution, and a possible remedy in this regard is to perform a logarithmic regression analysis.

Logarithmic regression means using the natural logarithm of a variable rather than the variable itself in a regression. Most often in practice, however, we replace only y with the log of y. When we want the logarithm of a number, we crunch it through a formula on a calculator (*Ln*) to get what we want. We call this 'logging.' The natural logarithm of, say, 5.0 is 1.605. More examples appear below.

Number on natural scale:	*Number on the logarithmic scale, i.e., in logs:*
0	Not defined
0.50	−0.693
1.00	0
8.40	2.128
10.00	2.302
50.00	3.912
100.55	4.611
500.00	6.215
5,000.77	8.517
100,000.00	11.519

The key feature to note is that logging makes small natural numbers change a little, say from 10.00 on the natural scale to 2.302 on the logarithmic scale. For large natural numbers, in contrast, the change is huge, such as from 100,000.00 (natural) to 11.519 (logarithmic). On the logarithmic scale, thus, larger natural numbers end up much closer to smaller natural numbers. This trivial fact has some favorable consequences when dealing with non-linearity, as we shall see below.

Practically, we first tell Stata or SPSS or our preferred software to create a logarithmic version of the variable of interest; see Section 6.2 on how to do this. We then use the logarithmic variable, and not the original variable, in the regression in the usual way. Figure 4.8 presents the plot and the regression line for the natural logarithm of bottle price (i.e., price measured in logs) and quality. Nothing changes in Stata or SPSS except for the new name given to the dependent variable.

The first thing to note is the y-axis. The values for the price variable on the log scale range from 1.9 to 5.5. This is much less variation than the original price scale from 7 to 252 Euros. The second thing to note is the closer fit between the regression line and the individual data points, that is, the red wines. The final thing to note is that there is no need for a non-linear regression line; the linear model specification fits nicely. That is, logging the price variable makes the non-linear association between bottle price and taste quality become linear in the present case. Yet the reason for bringing attention to this example is of course that this logging procedure works well in many cases involving y-variables having only positive values.[13] The regression results yielding Figure 4.8 appears in Stata–output 4.9.

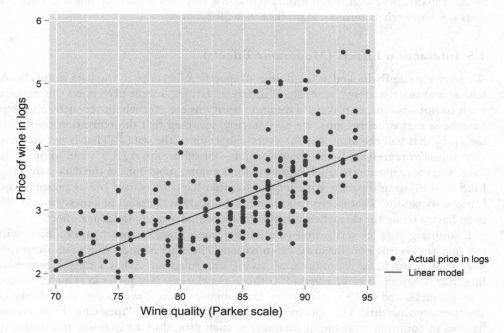

Figure 4.8 Scatterplot of correlation between taste quality and the natural logarithm of price of red wine, with regression line based on a linear model specification.

```
. reg log_price quality

      Source |       SS           df       MS            Number of obs   =       218
-------------+----------------------------------         F(1, 216)       =    150.17
       Model |  43.2566125          1  43.2566125        Prob > F        =    0.0000
    Residual |   62.220837        216  .288059431        R-squared       =    0.4101
-------------+----------------------------------         Adj R-squared   =    0.4074
       Total |   105.47745        217  .486071196        Root MSE        =    .53671


------------------------------------------------------------------------------
   log_price |      Coef.   Std. Err.      t    P>|t|     [95% Conf. Interval]
-------------+----------------------------------------------------------------
     quality |   .0751261   .0061306    12.25   0.000     .0630425    .0872096
       _cons |  -3.174961   .5186952    -6.12   0.000    -4.197313   -2.152609
------------------------------------------------------------------------------
```

Stata-output 4.9 Results producing Figure 4.8.

Stata–output 4.9 has one feature separating it from all regression outputs so far. The interpretation of the regression coefficient is *not* the average change in y given a one–unit increase in x. When y is measured in logs, the interpretation is instead the average change in y in *percent* given a one–unit increase in x.[14] A red wine with a taste quality score of, say, 90 thus costs 7.5 percent more on average than red wine with a taste quality score of 89.

One caveat regarding a regression where y is in logs and x is in natural levels: When the coefficient for such a regression is large, say ± 0.20 or larger, we should adjust it to get the correct percentage interpretation. We apply the formula $100 \times (e^b - 1)$ in this regard. A coefficient of, say, 0.45 thus becomes a 57 percent difference; $100 \times (2.7182^{0.45} - 1) = 56.83$. Finally, note that logging only works for variables with only positive values. It does not work when zero is a legitimate value for x or y.[15]

4.5 Interaction Effects (Moderator Effects)

We have repeatedly looked at the doings of specific subgroups in the data in this book: female and male students, students of different statuses, soccer players on the national team or not, and soccer players in different positions etc. Yet whenever such subgroups have been part of regressions, we have (tacitly) assumed that the regression coefficient for x_1, b_1, has had the same size for every subgroup in the data.[16] That is, we have assumed *parallel* regression lines – linear or not – for subgroups A, B, C, and so on in the data. Yet, presupposing parallel regression lines among subgroups in the data is the same kind of over-simplification with respect to what happens in real life as presupposing linear associations. That is, regression lines are not always parallel in a messy world. We even have a name for this when it happens: an interaction or moderator effect.

Examining parallel versus non–parallel regression lines has much in common with the linearity versus non–linearity examination. The start is often some cue from theory, prior research, or common sense saying that subgroup A has a steeper regression line than subgroup B or vice versa.[17] If this is not the case, we prefer keeping regression lines parallel based on Occam's razor.[18] But suppose we have such a cue, how do we go about examining this? The upcoming example shows that the "mechanical" procedure has lots in common with the quadratic regression procedure to examine non–linearity. Yet a difference is that non–parallel regression lines – that is, interaction effects – are not something we usually get a visual impression of by simply looking at a scatterplot as in the case of non–linearity.

We return to the student tourism data from Section 4.3. To set the stage, we look at how destination and length of stay are associated with total trip expenditures in a plain vanilla regression. The results appear in Stata-output 4.10. *Ceteris paribus*, trips to destinations beyond the Nordic countries on average incur 437 Euros more in expenditures than trips to the Nordic countries. A trip lasting for, say, ten days incurs 29 Euros more in expenditures on average than a trip lasting nine days *ceteris paribus*. Combined, the two x-variables explain 34.8 percent of the variation in total trip expenditures (R^2). Figure 4.9 presents the regression results of Stata-output 4.10 in a visual form.

```
. reg tot_spend i.destin los

      Source |       SS           df       MS            Number of obs   =     444
-------------+----------------------------------         F(2, 441)       =  117.92
       Model |  55052491.4          2   27526245.7       Prob > F        =  0.0000
    Residual |   102940870        441   233426.008       R-squared       =  0.3484
-------------+----------------------------------         Adj R-squared   =  0.3455
       Total |   157993361        443   356644.156       Root MSE        =  483.14


   tot_spend |      Coef.   Std. Err.        t    P>|t|     [95% Conf. Interval]
-------------+----------------------------------------------------------------
      destin |
Beyond Nordic |   436.8906   47.97093      9.11    0.000     342.6106    531.1707
         los |    28.9692   2.758541     10.50    0.000     23.54768    34.39072
       _cons |   32.28731   43.38848      0.74    0.457    -52.98658    117.5612
------------------------------------------------------------------------------
```

Stata-output 4.10 Total trip expenditures by trip destination and length of stay for the student tourism data.

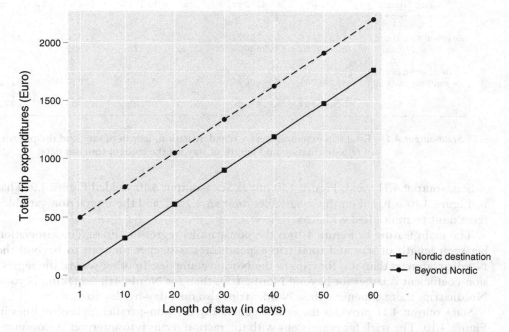

Figure 4.9 Graphical display of the multiple regression results in Stata-output 4.10. The parallel regression lines are a mathematical consequence of the linear model specification.

Three features are noteworthy regarding Figure 4.9. First, we note the upward-sloping regression lines for length of stay in step with the positive regression coefficient. Second, the regression line for the costlier non–Nordic trips is on top, whereas the similar line for the less expensive Nordic trips is at the bottom. Third, and this is the key point, the regression lines are parallel. Yet as mentioned above, this parallelism is a direct consequence of the inherent mathematical properties of the linear regression model. It does not necessarily mean that the two regression lines are parallel in real life.

Suppose a researcher has found out that length of stay's effect on trip expenditures is dependent on type of destination, that is, the effect is larger for some destinations and smaller for others. Such a scenario amounts to non–parallel regression lines in regression-speak. How do we examine this? We proceed as follows: First, we create a new variable, as we did for the quadratic regression. The new variable is the *product* of trip destination and length of stay. That is, we multiply `destin` and `los`. Second, we add the new product variable to the regression in Stata–output 4.10. Again, Stata has a built-in command that does this procedure in one step. (Section 4.9 shows the similar procedure in SPSS.) The results appear in Stata–output 4.11. The product variable, which we call an *interaction term*, [19] appears near the bottom of the output with the technical Stata name `destin#c.los`.[20]

```
. reg tot_spend i.destin##c.los

      Source |       SS           df       MS            Number of obs   =        444
-------------+-----------------------------------        F(3, 440)       =      95.41
       Model |  62271575.3         3   20757191.8        Prob > F        =     0.0000
    Residual |  95721785.7       440   217549.513        R-squared       =     0.3941
-------------+-----------------------------------        Adj R-squared   =     0.3900
       Total |   157993361       443   356644.156        Root MSE        =     466.42


-------------------------------------------------------------------------------------
   tot_spend |      Coef.   Std. Err.      t    P>|t|     [95% Conf. Interval]
-------------+-----------------------------------------------------------------------
      destin |
Beyond Nordic |   140.7215   69.19572     2.03   0.043     4.726339    276.7167
         los |   9.780752   4.264705     2.29   0.022     1.399028    18.16248
             |
destin#c.los |
Beyond Nordic |   31.45301   5.460095     5.76   0.000      20.7219    42.18411
             |
       _cons |    189.294   49.97386     3.79   0.000     91.07683    287.5111
-------------------------------------------------------------------------------------
```

Stata-output 4.11 Total trip expenditures by trip destination, length of stay, and the product of trip destination and length of stay for the student tourism data.

Stata–output 4.11 yields Figure 4.10 just as Stata–output 4.10 yielded Figure 4.9. That is, Figure 4.10 is based on the *x*-variables `destin`, `los`, and the interaction variable/ term `destin` multiplied with `los`.

The main feature of Figure 4.10 is the non–parallel regression lines. The association between length of stay and total trip expenditures is steeper for trips to beyond the Nordic countries than it is for trips to the Nordic countries. In other words, the regression coefficient is larger for beyond Nordic trips than for Nordic trips, making beyond Nordic trips more expensive than Nordic trips also on a day-by-day basis.

Stata–output 4.11 provides the results generating the non–parallel regression lines in Figure 4.10. The trick for regressions with interaction terms is to interpret the output – which is tricky! The first thing to note is that the regression coefficient for the interaction term is clearly different from zero: 31.45. Such a non-zero coefficient is the first

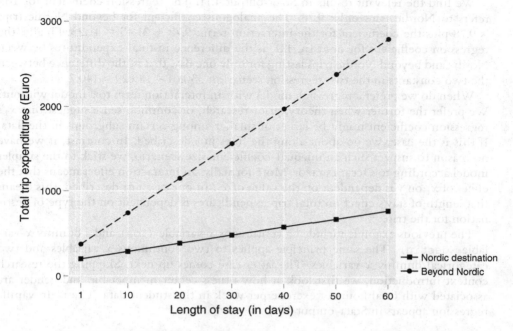

Figure 4.10 Graphical display of the multiple regression results in Stata-output 4.11. The non-parallel regression lines are a mathematical consequence of the linear model specification that includes an interaction variable/term.

cue suggesting that non-parallel regression lines fit the data better than parallel lines. The second cue is R^2. The plain vanilla regression had an R^2 of 0.348, cf. Stata-output 4.10, whereas the regression with the interaction term according to Stata-output 4.11 has an R^2 of 0.394. This is two strikes for non-parallel regression lines.[21] Now to the tricky part. A compact way of presenting the regression equation in Stata-output 4.11, in which total trip expenditures is TTE and omitting the error term, is

TTE = 189.29 + 140.72destin + 9.78los + 31.45destin×los.

For a trip to the Nordic countries, that is, destin = zero, this equation thus becomes

TTE = 189.29 + 140.72×0 + 9.78los + 31.45×0×los   →
TTE = 189.29 + 0 + 9.78los + 0   →
TTE = 189.29 + 9.78los.

That is, the regression coefficient for length of stay is 9.78 for trips to the Nordic countries. For trips to beyond the Nordic countries, that is, destin = one, the analogous equation becomes

TTE = 189.29 + 140.72×1 + 9.78los + 31.45×1×los   →
TTE = 189.29 + 140.72 + 9.78los + 31.45los   →
TTE = 330.01 + 41.23los.

For trips to beyond the Nordic countries, the regression coefficient for length of stay is 41.23. This much larger regression coefficient for the non-Nordic trips suggests a steeper regression line for this subgroup, as we saw in Figure 4.10.

We find the relevant results in Stata–output 4.11. The regression coefficient for los refers to Nordic trips only: 9.78. The analogous coefficient for beyond Nordic trips is 9.78 plus the coefficient for the interaction term: 9.78 + 31.45 = 41.23. Finally, the regression coefficient for destin, 141, is the difference in total expenditures between Nordic and beyond Nordic trips lasting for only one day, that is, the difference between the two constants in the two regression scenarios: 330.01 − 189.29 ≈ 140.7.

When do we prefer a regression model with an interaction term to a model without? We prefer the former when theory, prior research, or common sense suggest that x_1's regression coefficient might be larger or smaller among certain subgroups in the data. If this is the case, we go about along the lines just described. In contrast, if we have no reason to suspect such an unequal-coefficient-size scenario, we stick to the simpler model according to Occam's razor. More formally, an interaction effect means that the effect of x_1 on y is dependent on the value of x_2. In our case just described, this means that length of stay's effect on total trip expenditures is dependent on the type of destination for the trip.

The previous example included a continuous x-variable (los) and a dummy x-variable (destin). The same principle applies to two continuous x-variables and two categorical/dummy x-variables. The latter case comes up next. Skipping the research context introduction, we first look at how fitness center membership and gender are associated with total hours of exercise per week in the student data. The plain vanilla regression appears in Stata–output 4.12.

```
. reg hours_exer i.fitness_cen i.gender

      Source |       SS           df       MS          Number of obs   =        644
-------------+----------------------------------       F(2, 641)       =      67.60
       Model |  1109.64706          2  554.823531       Prob > F        =     0.0000
    Residual |  5261.01249        641  8.20750778       R-squared       =     0.1742
-------------+----------------------------------       Adj R-squared   =     0.1716
       Total |  6370.65955        643  9.90771314       Root MSE        =     2.8649


------------------------------------------------------------------------------
  hours_exer |      Coef.   Std. Err.      t    P>|t|     [95% Conf. Interval]
-------------+----------------------------------------------------------------
 fitness_cen |
         yes |   2.232115   .2259013     9.88   0.000     1.788519    2.675711
             |
      gender |
        male |   1.377412   .2361713     5.83   0.000     .9136494    1.841175
       _cons |   3.162703   .1795641    17.61   0.000     2.810098    3.515308
------------------------------------------------------------------------------
```

Stata-output 4.12 Results for multiple regression of hours of exercise by fitness center membership and student gender in the student exercise data.

The taken-for-granted assumption in Stata–output 4.12 is that the regression coefficient for the center membership dummy (i.e., 2.23 hours) has the same size for female and male students. Suppose, however, that we have come across research questioning this assumed gender-neutral effect, and that we wish to examine this more closely. Stata output 4.13 sheds light on the matter, following the exact same procedure as in Stata–output 4.11/Figure 4.10.

The regression coefficient for the fitness center dummy refers to the female students, that is, gender = zero. That is, female fitness center members exercise 1.78 more hours per week than female non-members on average. For the male students, that is, gender = one, the analogous difference is 3.06 (1.78 + 1.29 = 3.06). That is, the fitness center effect on exercise hours is larger among male students. The coefficient for gender, that is, 0.71, now refers to the gender difference in exercise hours for the non-members of fitness centers.

```
. reg hours_exer i.fitness_cen##i.gender

      Source |       SS           df       MS            Number of obs   =       644
-------------+----------------------------------         F(3, 640)       =     48.02
       Model |  1170.60467          3  390.201555         Prob > F        =    0.0000
    Residual |  5200.05488        640  8.12508576         R-squared       =    0.1837
-------------+----------------------------------         Adj R-squared   =    0.1799
       Total |  6370.65955        643  9.90771314         Root MSE        =    2.8505


------------------------------------------------------------------------------------
   hours_exer |      Coef.   Std. Err.      t    P>|t|     [95% Conf. Interval]
-------------+----------------------------------------------------------------------
  fitness_cen |
         yes |   1.776953   .2795229     6.36   0.000     1.22806    2.325846
             |
       gender |
        male |   .7137566   .3375251     2.11   0.035     .0509662   1.376547
             |
fitness_cen#gender |
     yes#male |   1.287862   .4701855     2.74   0.006     .3645689   2.211154
             |
        _cons |   3.388095      .1967    17.22   0.000     3.00184    3.774351
------------------------------------------------------------------------------------
```

Stata-output 4.13 Results for multiple regression of hours of exercise by fitness center membership, student gender, and the product of fitness center membership and student gender in the student exercise data.

Figure 4.11 shows the fitness center and gender interaction effect. We note the steeper regression line for male students in sync with the regression results in Stata-output 4.13.[22]

4.5.1 Non–Linearity and Interaction

We sometimes face situations in which interaction and non–linearity appear simultaneously. Figure 4.12 illustrates such a scenario for the student data. The age–exercise hours regression lines for the two genders are both non–linear (especially for males) and non–parallel. In other words, the non–linear age effect is dependent on the students' gender.

4.6 Regression on Experimental Data

Amos Tversky and Daniel Kahneman, two Israeli-born psychologists, went on a mission in the 1970s. In perfect hindsight, their ultimate target was the rational decision-maker at the core of the statistical models used by economists. In the early days, however, their aims were more modest: They wanted to show that most people have limited

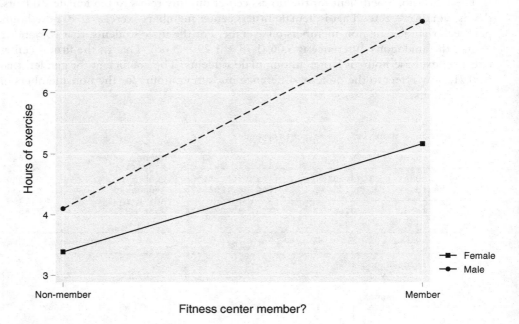

Figure 4.11 Graphical display of the multiple regression results in Stata–output 4.13. The non-parallel regression lines are a consequence of the linear model specification including an interaction term.

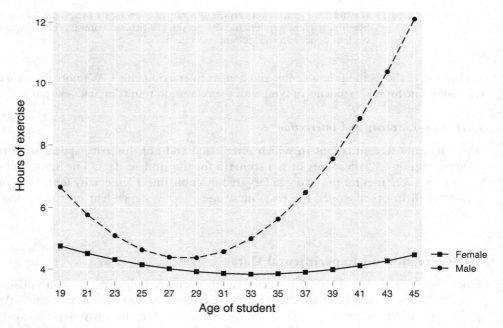

Figure 4.12 Hours of weekly exercise by gender and age. Non-linear regression model with interaction effect.

capabilities when it comes to doing assessments and making decisions based on numbers – in stark contrast to the 100 percent rational *homo economicus*. That is, people are subjected to a number of heuristics and biases. Using experiments with an RCT bent, the two psychologists forcefully showed their arguments rather than just mocking the flawed rationality assumptions of *homo economicus*. One of their first experiments, that set the scene many followed, has the following background.[23]

When people are about to do an assessment about making a decision or a judgment, they tend to rely on the *last* information they received. We call this the *anchoring effect*; the last information serves as an anchor for the assessment. This is not necessarily bad. What makes it a potentially distorting bias, however, is that we tend to rely on useless information as such anchors. This is irrational, to borrow a phrase from the economists' dictionary. Let us proceed with an update of the classical experiment.

Imagine a group of students receiving an assignment on a sheet of paper and being told not to look at each other while completing it. On top of the sheet is a short instruction on what to do followed by the question, 'What percentage of African nations do you think are members in the UN?' That is, the students' assignment is to make an educated guess as in writing down a number between 0 and 100. This is the experiment's y-variable.

The key to the experiment lies in the introduction *before* the question above. In this regard, one introduction might read:

> Read the instruction carefully, and do not show your answers to your 'neighbor!'
> A *random number* generator has chosen a number between 0 and 100.
> The number chosen and assigned to you is $X = 65$.

A *second* introduction may read like this:

> Read the instruction carefully, and do not show your answers to your 'neighbor!'
> A *random number* generator has chosen a number between 0 and 100.
> The number chosen and assigned to you is $X = 10$.

The only difference between the two introductions is the numbers at the end: 65 or 10. Whether a student gets a '65' or a '10' is the result of a coin toss (randomization) as in an RCT.[24] Furthermore, and provided the students follow the instruction, they do not get to know there are only two possible 'treatments:' 65 or 10. Neither does the 'doctor' (instructor) get to know which student gets which version of the introduction. The experiment is therefore double-blind. The treatment dummy is the x-variable of the experiment, where '65' is coded one (treatment) and '10' is coded zero (placebo).

Now for the logic in the experiment. The introduction with the bogus random generator is the anchor of interest. By definition, a *random* generator (fake or not) cannot provide any help in making an improved educated guess on the UN-percentage question. The anchor dummy should thus *not* affect the y-variable. That is, students receiving '65' (treatment) and students receiving '10' (placebo) should on average come up with about the same percentage guess on the UN question if they are rational. Is this really the case?

I have conducted the above experiment on some of my bachelor-level classes (N = 138). The data are called anchor_exp. If the two experiment groups come up with the same average guess on the UN question, the regression coefficient for the anchor dummy should be zero or thereabouts. Stata-output 4.14 presents the results in this regard.

```
. reg UN_percent i.anchor

      Source |       SS           df       MS       Number of obs   =        138
-------------+------------------------------        F(1, 136)       =      26.36
       Model |  15444.0478          1  15444.0478  Prob > F        =     0.0000
    Residual |  79667.6695        136  585.791688  R-squared       =     0.1624
-------------+------------------------------        Adj R-squared   =     0.1562
       Total |  95111.7174        137  694.246112  Root MSE        =     24.203


------------------------------------------------------------------------------
  UN_percent |      Coef.   Std. Err.      t    P>|t|     [95% Conf. Interval]
-------------+----------------------------------------------------------------
      anchor |
         65  |   21.19347   4.127557     5.13   0.000     13.03097    29.35596
       _cons |   24.82192   2.832763     8.76   0.000     19.21996    30.42388
------------------------------------------------------------------------------
```

Stata-output 4.14 Answer to UN question by anchor dummy in the anchor experiment data.

The constant is the mean UN-percentage guess for the placebo group, that is, the students receiving the 10-anchor. The average UN-guess for this 10-anchor group is almost 25 percent. For the students receiving the 65-anchor, that is, the treatment group, the average UN-guess is 21 percentage points higher – or 46 percent to be precise. This is a large difference illustrating that a nonsense anchor has a substantial effect on a later assessment.[25] Students, like people in general, are not rational when it comes to assessments having to do with numbers. This questions the assumptions of *homo economicus*.

Interesting as this experiment might be, the reason for bringing it up has to do with the distinction between experimental and statistical control mentioned in Sections 3.6, 3.7, and 4.2. I said there that experimental control makes statistical control redundant provided successful randomization. Now it is time to show this. Stata–output 4.15 presents two more regressions. The first adds the control variable gender to the model; the second also adds the control variable to class belonging (i.e., the constant = business administration class 1).

The uppermost multiple regression in Stata–output 4.15 shows roughly the same treatment effect (21.38) as the bivariate model in Stata–output 4.14. We note in passing that female students on average provide an almost 9 percentage points lower guess on the UN question than male students. The multiple regression model at the bottom of Stata–output 4.15 shows about the same treatment effect (21.65) as the upper model. The point now is that we could go on and on adding more x-variables to the regression model without anything of much interest happening to the treatment effect; it will probably remain in the 21–22-point region. The reason is that the additional control variables will have no systematic association with the treatment dummy provided successful randomization.[26] One caution is in order, however. The treatment effect is the average effect among all the units in the data. There could be interaction effects lurking in the background, and we should examine these along the lines already drawn up in Section 4.5.

```
. reg UN_percent i.anchor i.gender
```

Source	SS	df	MS
Model	17940.945	2	8970.47248
Residual	77170.7724	135	571.635351
Total	95111.7174	137	694.246112

Number of obs =	138
F(2, 135) =	15.69
Prob > F =	0.0000
R-squared =	0.1886
Adj R-squared =	0.1766
Root MSE =	23.909

UN_percent	Coef.	Std. Err.	t	P>|t|	[95% Conf. Interval]	
anchor						
65	21.37796	4.078333	5.24	0.000	13.31227	29.44365
gender						
female	-8.582597	4.106558	-2.09	0.038	-16.7041	-.4610894
_cons	28.46658	3.297232	8.63	0.000	21.94567	34.98749

```
. reg UN_percent i.anchor i.gender i.class
```

Source	SS	df	MS
Model	19483.748	4	4870.93699
Residual	75627.9694	133	568.631349
Total	95111.7174	137	694.246112

Number of obs =	138
F(4, 133) =	8.57
Prob > F =	0.0000
R-squared =	0.2049
Adj R-squared =	0.1809
Root MSE =	23.846

UN_percent	Coef.	Std. Err.	t	P>|t|	[95% Conf. Interval]	
anchor						
65	21.64521	4.072587	5.31	0.000	13.58979	29.70063
gender						
female	-8.722002	4.303204	-2.03	0.045	-17.23357	-.2104307
class						
busi_adm_2	7.814073	5.086741	1.54	0.127	-2.247303	17.87545
mixed	5.196702	5.032269	1.03	0.304	-4.756931	15.15033
_cons	25.13942	3.859401	6.51	0.000	17.50567	32.77316

Stata-output 4.15 Answer to UN-question by anchor dummy and controls in the anchor experiment data.

4.7 A Dummy y

We have analyzed many dummy variables in this book, but they have exclusively been x-variables save for in Section 3.2. In real life, however, such dummies are often y-variables. Some examples include (the decision) to vote or not, being employed or not, or being ill or not. That is, we examine y-variables that directly or indirectly may be coded as one (the presence of something/yes) or zero (the non-presence of something/ no). Because there is an abundance of such 'choice' y-variables in the behavioral and social sciences, it is mandatory that a quantitative analyst worth her salt manages to handle this frequently occurring situation within a multiple regression framework. This section teaches you how to do so.[27]

The good news at the outset is that we often may keep on doing plain vanilla multiple regression for a dummy y the way we have done so far.[28] The next example follows up the analysis in Section 3.2. The dependent variable is whether a student is member of a sports club or not, and we examine how certain x-variables are possibly associated with this yes/no choice. Stata-output 4.16 shows summary descriptive statistics for the sports club membership variable and breaks it down on gender by means of a cross-tabulation.

```
. tab sport_club

sport_club |      Freq.       Percent         Cum.
-----------+-----------------------------------------
        no |        504         78.26        78.26
       yes |        140         21.74       100.00
-----------+-----------------------------------------
     Total |        644        100.00

. sum sport_club

    Variable |        Obs         Mean    Std. Dev.         Min          Max
-------------+-----------------------------------------------------------------
  sport_club |        644     .2173913     .4127916           0            1

. tab sport_club gender, col

           |       gender
sport_club |    female        male |      Total
-----------+------------------------+-----------
        no |       356         148 |        504
           |     85.58       64.91 |      78.26
-----------+------------------------+-----------
       yes |        60          80 |        140
           |     14.42       35.09 |      21.74
-----------+------------------------+-----------
     Total |       416         228 |        644
           |    100.00      100.00 |     100.00
```

Stata-output 4.16 Descriptive statistics for sport club membership variable and sports club membership by gender.

We see on top of the output that presently 21.74 percent of the students are sports club members. This corresponds to the mean of 0.2174 in the middle of the output. Finally, the lower part of the output suggests that male students are sports club members more often than female students. The gender difference is about 21 percentage points ($35 - 14 = 21$). Stata-output 4.17 shows a bivariate regression between the gender dummy and the membership dummy in the usual manner.

The constant refers to the mean of y for x = zero. That is, the probability of being a sports club member among females is roughly 14 percent according to Stata-output 4.17.

```
. reg sport_club i.gender

      Source |       SS           df       MS            Number of obs   =      644
-------------+----------------------------------            F(1, 642)       =    39.10
       Model |  6.28923898         1  6.28923898           Prob > F        =   0.0000
    Residual |  103.275978       642   .16086601           R-squared       =   0.0574
-------------+----------------------------------            Adj R-squared   =   0.0559
       Total |  109.565217       643   .170396917          Root MSE        =   .40108

------------------------------------------------------------------------------
  sport_club |      Coef.   Std. Err.       t     P>|t|    [95% Conf. Interval]
-------------+----------------------------------------------------------------
    gender |
      male |   .2066464   .0330492     6.25    0.000    .1417488    .271544
     _cons |   .1442308   .0196646     7.33    0.000     .105616   .1828455
------------------------------------------------------------------------------
```

Stata-output 4.17 Results for regression of sports club membership by gender in the student
exercise data.

This mirrors the result in the cross-table in Stata-output 4.16. The regression coefficient
0.207 implies that the probability of being a sports club member among males is almost
21 percentage points larger than 14 percent: 35 percent. This result also mirrors the
cross-table in Stata-output 4.16.[29] We may thus deduce the general interpretation of the
linear regression coefficient when y is a dummy: the average probability change for $y =$
1 in percentage points given a one-unit increase in x.

We extend the bivariate regression model into a multiple regression model as before.
The multiple regression in Stata-output 4.18 adds the youth sports involvement variable
and the exercise preference variable in the customary manner.

```
. reg sport_club i.gender youth_exe i.exer_most

      Source |       SS           df       MS            Number of obs   =      644
-------------+----------------------------------            F(4, 639)       =    22.37
       Model |  13.458707          4  3.36467675          Prob > F        =   0.0000
    Residual |  96.1065104       639   .150401425         R-squared       =   0.1228
-------------+----------------------------------            Adj R-squared   =   0.1173
       Total |  109.565217       643   .170396917          Root MSE        =   .38782

------------------------------------------------------------------------------
  sport_club |      Coef.   Std. Err.       t     P>|t|    [95% Conf. Interval]
-------------+----------------------------------------------------------------
    gender |
      male |   .1758789   .0330256     5.33    0.000    .1110272   .2407306
 youth_exe |   .0326773   .0054216     6.03    0.000     .022031   .0433236
           |
  exer_most |
    cardio |    .112308   .0416231     2.70    0.007    .0305734   .1940426
     equal |   .0679021    .035332     1.92    0.055   -.0014788    .137283
           |
     _cons |  -.1139902   .0426462    -2.67    0.008    -.197734  -.0302465
------------------------------------------------------------------------------
```

Stata-output 4.18 Results for linear regression of sports club membership by gender, youth
sports involvement, and exercise preference in the student exercise data.

The following *ceteris paribus* interpretations apply: A male student has on average a 17.59 percentage points greater probability of being a sports club member than a female student. A student with the score of, say, six on the youth sports variable has on average a 3.27 percentage points greater probability of being a sports club member than a student with the score of five. Finally, a student preferring cardio training has on average an 11.23 percentage points greater probability of being a sports club member than a student preferring strength training (i.e., the reference). A similar difference between the 'preferring both equally much' group and the 'strength training' group is 6.79 percentage points.

The direct extension from a linear regression of a continuous y and the translucent interpretation of the regression coefficient explain why linear regression for a dummy y has become popular in recent years. We call it the Linear Probability Model (LPM). For many years, however, the LPM was a no-go in statistics circles.[30] The reason is that it violates some of the assumptions of regression analysis; see Section 6.6 for more on this. The appropriate statistical model was – and, in many fields of research, still is – logistic regression analysis. We therefore cover this model below.

4.7.1 Logistic Regression Analysis

The equivalent logistic regression of the model in Stata-output 4.18 appears in Stata-output 4.19.

```
. logit sport_club i.gender youth_exe i.exer_most

Output omitted

Logistic regression                             Number of obs    =       644
                                                LR chi2(4)       =     87.61
                                                Prob > chi2      =    0.0000
Log likelihood = -293.38478                     Pseudo R2        =    0.1299

------------------------------------------------------------------------------
 sport_club |      Coef.   Std. Err.      z    P>|z|     [95% Conf. Interval]
------------+-----------------------------------------------------------------
     gender |
       male |   1.061623    .2107988     5.04   0.000     .648465    1.474781
  youth_exe |   .2640826    .0454961     5.80   0.000     .1749119    .3532533
            |
   exer_most |
     cardio |   .8055775    .2832373     2.84   0.004     .2504426    1.360712
      equal |   .5009095    .2483427     2.02   0.044     .0141667    .9876522
            |
      _cons |  -4.067698    .4155778    -9.79   0.000    -4.882216   -3.253181
------------------------------------------------------------------------------
```

Stata-output 4.19 Results for logistic regression of sports club membership by gender, youth sports involvement, and exercise preference in the student exercise data.

The logistic regression coefficient for gender is 1.062, and the analogous youth exercise coefficient is 0.264. Unfortunately, however, these coefficients tell us nothing except for their signs. The positive gender coefficient implies that male students are *more* likely to be members of sports clubs than female students – which we already know from Stata-output 4.18. The positive youth exercise coefficient similarly implies that a student with a score of six on this variable is *more* likely to be a sports club member than a student with a score of five – which we also know from Stata-output 4.18. The upshot is that although the logistic regression model solves some of the problems with the LPM model, such a rescue mission comes at a steep price: non-informative logistic regression coefficients.[31] SPSS-output 4.2 presents a similar output for the logistic regression in SPSS format. Column B displays the logistic regression coefficients.

Variables in the Equation

		B	S.E.	Wald	df	Sig.	Exp(B)
Step 1[a]	gender(1)	1,062	,211	25,363	1	,000	2,891
	youth_exe	,264	,045	33,692	1	,000	1,302
	cardio(1)	,806	,283	8,089	1	,004	2,238
	equal(1)	,501	,248	4,068	1	,044	1,650
	Constant	-4,068	,416	95,806	1	,000	,017

a. Variable(s) entered on step 1: gender, youth_exe, cardio, equal.

SPSS-*output 4.2* Results for logistic regression of sports club membership by gender, youth sports involvement, and exercise preference in the student exercise data.

By default, SPSS also reports so-called *odds-ratios* in the column on the right: Exp(B). Just as any student has a probability of being a sports club member, she also has the odds of being a member of such a club.[32] An odds-ratio is one such odds divided by another, as in for example male students' odds divided by female students' odds. According to SPSS-output 4.2, male students' odds of being sports club members are 2.89 times larger than female students' odds of being sports club members *ceteris paribus*. (An odds-ratio of 1.0 would suggest gender parity in this respect.) A student scoring six on the youth sports variable has 1.30 greater odds of being a sports club member than a student scoring five *ceteris paribus*. Similar interpretations hold for the exercise preference dummies.[33] One caveat is in order regarding odds-ratios. They say little of the magnitude of an effect. Moreover, an odds-ratio of 2.0 most often does *not* mean twice as likely in terms of probabilities. An efficient solution to non-informative logistic coefficients and odds-ratios is to turn logistic regression coefficients into something resembling linear regression coefficients. Please welcome *marginal effects*!

4.7.2 Logistic Regression: Marginal Effects

Marginal effects based on logistic regression analysis have roughly the same interpretation as the regression coefficients of the LPM. Stata-output 4.20 presents the marginal effects for the logistic regression model in Stata-output 4.19.

The *ceteris paribus* interpretations are straightforward: A male student has on average a 16.92 percentage points larger probability of being a sports club member than a female student. A student with a score of, say, six on the youth sports variable has on average a

```
. margins, dydx(*)

Average marginal effects                      Number of obs    =        644
Model VCE     : OIM

Expression    : Pr(sport_club), predict()
dy/dx w.r.t.  : 1.gender youth_exe 1.exer_most 2.exer_most
```

```
-----------------------------------------------------------------------------
             |            Delta-method
             |    dy/dx   Std. Err.      z    P>|z|    [95% Conf. Interval]
-------------+---------------------------------------------------------------
    gender   |
      male   |  .1692471   .0344013    4.92   0.000    .1018217    .2366724
 youth_exe   |  .0389276   .0062707    6.21   0.000    .0266372     .051218
             |
  exer_most  |
    cardio   |  .1184235   .0418769    2.83   0.005    .0363462    .2005008
     equal   |  .0688692     .03329    2.07   0.039     .003622    .1341165
-----------------------------------------------------------------------------
```

Note: dy/dx for factor levels is the discrete change from the base level.

Stata-output 4.20 Marginal effects based on logistic regression model in Stata-output 4.19.

3.89 percentage points greater probability of being a sports club member than a student with a score of five. Finally, a student preferring cardio training has on average an 11.84 percentage points greater probability of being a sports club member than a student preferring strength training (i.e., the reference). The similar difference between the 'preferring both equally much' group and the 'strength training' group is 6.89 percentage points. In general, these marginal effects are very close to the plain vanilla regression coefficients (i.e., LPM coefficients) in Stata-output 4.18.[34]

The similarity of marginal effects based on logistic regression and LPM coefficients appears to be the general case (cf. Angrist and Pischke, 2009). This most probably explains why many analysts prefer the LPM despite its shortcomings: It produces good estimates and we do not need a sophisticated statistics program to calculate it. In fact, Excel will suffice nicely.[35]

4.7.3 Logistic Regression: Pseudo R^2

When y is continuous or roughly so, it makes intuitive sense to talk about y's variation and how much of this variation the x-variables account for. In short, the idea of R^2 adds up. Such 'variation' makes less sense when y is a dummy. For this reason, and some technical ones we need not get into, logistic regression has no universal R^2 measure. Instead, we have a plethora of Pseudo R^2 measures. We note for Stata-output 4.19 that our Pseudo R^2 is almost 13 percent. Although this measure has a certain resemblance to the plain vanilla R^2, we should not interpret it literally. It is probably also fair to say that the various Pseudo R^2 measures play a smaller role in logistic regression than R^2 does in linear regression.

4.7.4 Logistic Regression: Non-Linearity and Interaction Effects

Non-linearity and interaction effects, as we saw in Sections 4.4 and 4.5, apply to the case of a dummy y in the same way as for a continuous y, save for the use of a logarithmic y. I thus skip showing this.

4.8 Chapter Summary, Key Learning Points, and Further Reading

This chapter has been about multiple regression analysis. A multiple regression model has one y-variable and multiple x-variables. Below follows some key learning points:

- A multiple regression model per definition has more than one x-variable. Such a model may be expressed in an algebraic or visual form.
- Multiple regression analysis is a typical way of holding constant (or controlling for) other x-variables when analyzing how x_1 affects y for observational data in the social and behavioral sciences.
- To find the causal or unique effect of x_1 on y when analyzing observational data, we invariably in practice have to control for several other x-variables.
- R^2 measures how much of the variation (variance) in y the x-variables account for combined.
- Non-linearity and interaction effects may – and should sometimes! – be built into multiple regression models. Theory, prior research, and common-sense guide when to check out non-linear regression models and regression models that include interaction terms.
- If carried out successfully, an experimental design/RCT typically makes statistical control for other x-variables redundant.
- Linear multiple regression most often handles a dummy y-variable very well. This is called the LPM.
- Logistic regression analysis is tailor-made for the dummy y-variable case, but it needs some tinkering to produce translucent regression coefficients.

Pearl and Mackenzie (2018) and Rosenbaum (2017) are still the key sources for doing causal analysis on observational or experimental data. Cunningham (2021) is a recent and easy-going contribution to this causal inference literature. Thrane (2020) provides an in-depth but practically oriented coverage of all things related to multiple regression analysis; see also Wolf and Best (2015). Long (1997, 2015) and Hosmer et al. (2013) teaches you all you need to know about logistic regression analysis and related techniques. Mitchell (2021) is the prime source for visualizing regression results in Stata.

Best and Wolf (2015) is the next natural step for all things related to regression and causal analysis in a more practical manner. The same goes for Angrist and Pischke (2009, 2015) and Berk (2004). Imbens and Rubin (2015) is a comprehensive primer to causal inference in experimental settings, but it requires a certain amount of competence in math/calculus.

4.9 Do-Files in Stata and Syntax-Files in SPSS

I assume you have read Sections 2.9 and 3.9 before taking on this section. The commands appear as usual in plain text 'outside' of do-files (Stata) or syntax-files (SPSS) to save space. As usual, I add some comments to the commands on occasion. I assume throughout that the 'correct' data set is in memory to avoid unnecessary repetition.

4.9.1 Stata-Commands in Do-Files

Table 4.1

```
sum hours_exer
```

Table 4.2

```
tab fitness_center
tab exer_easy_motive
```

Stata-output 4.1

```
reg hours_exer i.fitness_center
```

Stata-output 4.2

```
reg hours_exer i.fitness_center exer_easy_motive
```

To do a multiple regression in Stata we simply add more *x*-variables after the first. The variable `exer_easy_motive` thus makes the above regression a multiple regression.

Stata-output 4.3

```
reg hours_exer i.fitness_center exer_easy_motive exer_imp_qol
```

Figure 4.1

```
twoway (scatter price quality)(lfit price quality)
```

Stata-output 4.4

```
reg price quality
```

Stata-output 4.5

```
reg price quality alch_perc
```

Figure 4.2

```
reganat price quality alch_perc, dis(quality) biline
```

Before applying the reganat-command you must first download it. In the Command-window, type `findit reganat` and follow the instructions after first clicking on SJ-13-1 st0285.

Table 4.3

```
sum tot_spend los book_time destin type_trip
```

Stata-output 4.6

```
reg tot_spend los book_time i.destin i.type_trip
```

Figure 4.5

```
twoway (scatter inc_year age)(lfit inc_year age)
```

The l in lfit is short for linear (fit).

Figure 4.6

```
twoway (scatter inc_year age) (qfit inc_year age)
```

The q in qfit is short for quadratic (fit).

Stata output 4.7

```
reg inc_year c.age##c.age
```

The ## between the variable names means that both age and age × age (i.e. age2) should be included in the regression model. The c. tells Stata that age is, or should be treated as, a continuous variable.

Stata-output 4.8

```
reg inc_year i.age_ord
```

Figure 4.7

```
twoway (scatter price quality) (qfit price quality) (lfit price
quality)
```

Figure 4.8

```
twoway (scatter log_price quality) (lfit log_price quality)
```

Stata-output 4.9

```
reg log_price quality
```

Stata-output 4.10

```
reg tot_spend i.destin los
```

Figure 4.9
Two commands immediately after the regression command in Stata-output 4.10 are necessary to produce the graph:

```
margins destin, at(los=(1 10 20 30 40 50 60))
marginsplot, noci
```

Stata-output 4.11

```
reg tot_spend i.destin##c.los
```

The ## between the variable names means that both destin, los, and destin × los (i.e., the interaction term) should be included in the regression model. The c. tells Stata that los is, or should be treated as, a continuous variable.

Figure 4.10

```
margins destin, at(los=(1 10 20 30 40 50 60))
marginsplot, noci
```

Stata-output 4.12

```
reg hours_exer i.fitness_cen i.gender
```

Stata-output 4.13

```
reg hours_exer i.fitness_cen##i.gender
```

Figure 4.11

```
reg hours_exer i.fitness_cen##i.gender
margins gender, at(fitness_cen =(0(1)1))
marginsplot, noci
```

Figure 4.12
The sequence of commands to generate the graph are:

```
reg hours_exer i.gender##c.age##c.age i.sport_club i.fitness_cen
margins gender, at(age=(19(2)45))
marginsplot, noci
```

Stata-output 4.14

```
reg UN_percent i.anchor
```

Stata-output 4.15

```
reg UN_percent i.anchor i.gender
reg UN_percent i.anchor i.gender i.class
```

Stata-output 4.16

```
tab sport_club
sum sport_club
tab sport_club gender, col
```

Stata-output 4.17

```
reg sport_club i.gender
```

Stata-output 4.18

```
reg sport_club i.gender youth_exe i.exer_most
```

Stata-output 4.19

```
logit sport_club i.gender youth_exe i.exer_most
```

The only new feature is the replacement of reg with logit as the command. To obtain odds–ratios, use the command:

```
logit sport_club i.gender youth_exe i.exer_most, or
```

Stata-output 4.20
Immediately after the logistic regression in Stata-output 4.19, use the command:

```
margins, dydx (*)
```

4.9.2 SPSS-Commands in Syntax-Files

Table 4.1

```
DESCRIPTIVES VARIABLES=hours_exer
  /STATISTICS=MEAN STDDEV MIN MAX.
```

Table 4.2

```
FREQUENCIES VARIABLES=fitness_center exer_easy_motive
  /ORDER=ANALYSIS.
```

Stata-output 4.1

```
REGRESSION
  /MISSING LISTWISE
  /STATISTICS COEFF OUTS R ANOVA
  /CRITERIA=PIN(.05) POUT(.10)
  /NOORIGIN
  /DEPENDENT hours_exer
  /METHOD=ENTER fitness_center.
```

Stata-output 4.2

```
REGRESSION
  /MISSING LISTWISE
  /STATISTICS COEFF OUTS R ANOVA
  /CRITERIA=PIN(.05) POUT(.10)
  /NOORIGIN
  /DEPENDENT hours_exer
  /METHOD=ENTER fitness_center exer_easy_motive.
```

To do a multiple regression in SPSS we simply add more *x*-variables after the first. The variable exer_easy_motive thus makes the above regression a multiple regression.

Stata-output 4.3

```
REGRESSION
  /MISSING LISTWISE
  /STATISTICS COEFF OUTS R ANOVA
  /CRITERIA=PIN(.05) POUT(.10)
  /NOORIGIN
  /DEPENDENT hours_exer
  /METHOD=ENTER fitness_center exer_easy_motive exer_imp_qol.
```

Figure 4.1

```
GGRAPH
  /GRAPHDATASET NAME="graphdataset" VARIABLES=quality price MISS-
ING=LISTWISE REPORTMISSING=NO
  /GRAPHSPEC SOURCE=INLINE
  /FITLINE TOTAL=YES.
BEGIN GPL
  SOURCE: s=userSource(id("graphdataset"))
  DATA: quality=col(source(s), name("quality"))
  DATA: price=col(source(s), name("price"))
  GUIDE: axis(dim(1), label("quality"))
  GUIDE: axis(dim(2), label("price"))
  GUIDE: text.title(label("Simple Scatter with Fit Line of price by
quality"))
  ELEMENT: point(position(quality*price))
END GPL.
```

Stata–output 4.4

```
REGRESSION
  /DESCRIPTIVES MEAN STDDEV CORR SIG N
  /MISSING LISTWISE
  /STATISTICS COEFF OUTS R ANOVA
  /CRITERIA=PIN(.05) POUT(.10)
  /NOORIGIN
  /DEPENDENT price
  /METHOD=ENTER quality.
```

Stata–output 4.5

```
REGRESSION
  /DESCRIPTIVES MEAN STDDEV CORR SIG N
  /MISSING LISTWISE
  /STATISTICS COEFF OUTS R ANOVA
  /CRITERIA=PIN(.05) POUT(.10)
  /NOORIGIN
  /DEPENDENT price
  /METHOD=ENTER quality alch_perc.
```

Figure 4.2
To the best of my knowledge, there is no SPSS option readily available for generating this graph.

Table 4.3

```
DESCRIPTIVES VARIABLES=tot_spend los book_time destin type_trip
  /STATISTICS=MEAN STDDEV MIN MAX.
```

SPSS–output 4.1 (Stata–output 4.6)

```
REGRESSION
  /DESCRIPTIVES MEAN STDDEV CORR SIG N
  /MISSING LISTWISE
  /STATISTICS COEFF OUTS R ANOVA
  /CRITERIA=PIN(.05) POUT(.10)
  /NOORIGIN
  /DEPENDENT tot_spend
  /METHOD=ENTER los book_time destin type_trip.
```

Figure 4.5

```
GGRAPH
  /GRAPHDATASET NAME="graphdataset" VARIABLES=age inc_year MISS-
ING=LISTWISE REPORTMISSING=NO
  /GRAPHSPEC SOURCE=INLINE
  /FITLINE TOTAL=YES.
BEGIN GPL
  SOURCE: s=userSource(id("graphdataset"))
  DATA: age=col(source(s), name("age"))
  DATA: inc_year=col(source(s), name("inc_year"))
  GUIDE: axis(dim(1), label("age"))
  GUIDE: axis(dim(2), label("inc_year"))
  GUIDE: text.title(label("Simple Scatter with Fit Line of inc_year
by age"))
  ELEMENT: point(position(age*inc_year))
END GPL.
```

Figure 4.6

```
TSET NEWVAR=NONE.
CURVEFIT
  /VARIABLES=inc_year WITH age
  /CONSTANT
  /MODEL=QUADRATIC
  /PLOT FIT.
```

Stata–output 4.7

```
COMPUTE age_square=age*age.
REGRESSION
  /DESCRIPTIVES MEAN STDDEV CORR SIG N
  /MISSING LISTWISE
  /STATISTICS COEFF OUTS R ANOVA
  /CRITERIA=PIN(.05) POUT(.10)
  /NOORIGIN
  /DEPENDENT inc_year
  /METHOD=ENTER age age_square.
```

In SPSS, we must first create the variable age_square (i.e., age^2) by the COMPUTE-command before using it in the quadratic regression.

Stata–output 4.8

```
RECODE age_ord (2=1) (ELSE=0) INTO age21_23.
RECODE age_ord (3=1) (ELSE=0) INTO age24_26.
RECODE age_ord (4=1) (ELSE=0) INTO age27_29.
RECODE age_ord (5=1) (ELSE=0) INTO age30_32.
RECODE age_ord (6=1) (ELSE=0) INTO age33_41.
REGRESSION
  /DESCRIPTIVES MEAN STDDEV CORR SIG N
  /MISSING LISTWISE
  /STATISTICS COEFF OUTS R ANOVA
  /CRITERIA=PIN(.05) POUT(.10)
  /NOORIGIN
  /DEPENDENT inc_year
  /METHOD=ENTER age21_23 age24_26 age27_29 age30_32 age33_41.
```

In SPSS, we must first create the various age-group dummies by the RECODE-commands before using them in the regression.

Figure 4.7

```
TSET NEWVAR=NONE.
CURVEFIT
  /VARIABLES=price WITH quality
  /CONSTANT
  /MODEL=LINEAR QUADRATIC
  /PLOT FIT.
```

Figure 4.8

```
GGRAPH
  /GRAPHDATASET NAME="graphdataset" VARIABLES=quality log_price
MISSING=LISTWISE REPORTMISSING=NO
  /GRAPHSPEC SOURCE=INLINE
  /FITLINE TOTAL=YES.
BEGIN GPL
  SOURCE: s=userSource(id("graphdataset"))
  DATA: quality=col(source(s), name("quality"))
  DATA: log_price=col(source(s), name("log_price"))
  GUIDE: axis(dim(1), label("quality"))
  GUIDE: axis(dim(2), label("log_price"))
  GUIDE: text.title(label("Simple Scatter with Fit Line of log_price
by quality"))
  ELEMENT: point(position(quality*log_price))
END GPL.
```

Stata-output 4.9

```
REGRESSION
  /DESCRIPTIVES MEAN STDDEV CORR SIG N
  /MISSING LISTWISE
  /STATISTICS COEFF OUTS R ANOVA
  /CRITERIA=PIN(.05) POUT(.10)
  /NOORIGIN
  /DEPENDENT log_price
  /METHOD=ENTER quality.
```

Stata-output 4.10

```
REGRESSION
  /DESCRIPTIVES MEAN STDDEV CORR SIG N
  /MISSING LISTWISE
  /STATISTICS COEFF OUTS R ANOVA
  /CRITERIA=PIN(.05) POUT(.10)
  /NOORIGIN
  /DEPENDENT tot_spend
  /METHOD=ENTER destin los.
```

Figure 4.9

I know of no easy way to produce a graph like Figure 4.9 in SPSS. That said, Excel might be used to generate the *y*-values that, in turn, could go into a graph.

Stata-output 4.11

```
COMPUTE destin_los=los * destin.
REGRESSION
  /DESCRIPTIVES MEAN STDDEV CORR SIG N
  /MISSING LISTWISE
  /STATISTICS COEFF OUTS R ANOVA
  /CRITERIA=PIN(.05) POUT(.10)
  /NOORIGIN
  /DEPENDENT tot_spend
  /METHOD=ENTER destin los destin_los.
```

In SPSS, we must first create the product variable/term destin_los by the COMPUTE-command before using it in the interaction regression model.

Figure 4.10

```
GRAPH
  /SCATTERPLOT(BIVAR)=los WITH tot_spend BY destin
  /MISSING=LISTWISE.
```

Now, first double-click on the graph to get it 'active.' Then click on Elements and choose Fit Line at Subgroups. The appearing graph resembles Figure 4.10 although it also contains the individual data points.

Stata-output 4.12

```
REGRESSION
  /DESCRIPTIVES MEAN STDDEV CORR SIG N
  /MISSING LISTWISE
  /STATISTICS COEFF OUTS R ANOVA
  /CRITERIA=PIN(.05) POUT(.10)
  /NOORIGIN
  /DEPENDENT hours_exer
  /METHOD=ENTER fitness_cen gender.
```

Stata-output 4.13

```
COMPUTE fit_cen_gender=fitness_cen * gender.
REGRESSION
  /DESCRIPTIVES MEAN STDDEV CORR SIG N
  /MISSING LISTWISE
  /STATISTICS COEFF OUTS R ANOVA
  /CRITERIA=PIN(.05) POUT(.10)
  /NOORIGIN
  /DEPENDENT hours_exer
  /METHOD=ENTER fitness_cen gender fit_cen_gender.
```

Figure 4.11

```
GRAPH
  /SCATTERPLOT(BIVAR)=fitness_cen WITH hours_exer BY gender
  /MISSING=LISTWISE.
```

Now, first double-click on the graph to get it "active." Then click on Elements and choose Fit Line at Subgroups. The appearing graph resembles Figure 4.11 although it also contains the individual data points.

Figure 4.12
To the best of my knowledge there is no SPSS option readily available for generating this graph.

Stata-output 4.14

```
REGRESSION
  /DESCRIPTIVES MEAN STDDEV CORR SIG N
  /MISSING LISTWISE
  /STATISTICS COEFF OUTS R ANOVA
  /CRITERIA=PIN(.05) POUT(.10)
  /NOORIGIN
  /DEPENDENT UN_percent
  /METHOD=ENTER anchor.
```

Stata-output 4.15

```
REGRESSION
  /DESCRIPTIVES MEAN STDDEV CORR SIG N
  /MISSING LISTWISE
```

```
  /STATISTICS COEFF OUTS R ANOVA
  /CRITERIA=PIN(.05) POUT(.10)
  /NOORIGIN
  /DEPENDENT UN_percent
  /METHOD=ENTER anchor gender.
RECODE class (2=1) (ELSE=0) INTO busi_adm_2.
RECODE class (3=1) (ELSE=0) INTO mixed.
REGRESSION
  /DESCRIPTIVES MEAN STDDEV CORR SIG N
  /MISSING LISTWISE
  /STATISTICS COEFF OUTS R ANOVA
  /CRITERIA=PIN(.05) POUT(.10)
  /NOORIGIN
  /DEPENDENT UN_percent
  /METHOD=ENTER anchor gender busi_adm_2 mixed.
```

Stata-output 4.16

```
FREQUENCIES VARIABLES=sport_club
  /ORDER=ANALYSIS.
DESCRIPTIVES VARIABLES=sport_club
  /STATISTICS=MEAN STDDEV MIN MAX.
CROSSTABS
  /TABLES=sport_club BY gender
  /FORMAT=AVALUE TABLES
  /CELLS=COUNT COLUMN
  /COUNT ROUND CELL.
```

Stata-output 4.17

```
REGRESSION
  /DESCRIPTIVES MEAN STDDEV CORR SIG N
  /MISSING LISTWISE
  /STATISTICS COEFF OUTS R ANOVA
  /CRITERIA=PIN(.05) POUT(.10)
  /NOORIGIN
  /DEPENDENT sport_club
  /METHOD=ENTER gender.
```

Stata-output 4.18

```
RECODE exer_most (1=1) (ELSE=0) INTO cardio.
RECODE exer_most (2=1) (ELSE=0) INTO equal.
REGRESSION
  /DESCRIPTIVES MEAN STDDEV CORR SIG N
  /MISSING LISTWISE
  /STATISTICS COEFF OUTS R ANOVA
  /CRITERIA=PIN(.05) POUT(.10)
  /NOORIGIN
  /DEPENDENT sport_club
  /METHOD=ENTER gender youth_exe cardio equal.
```

SPSS–output 4.2 (Stata–output 4.19)

```
LOGISTIC REGRESSION VARIABLES sport_club
  /METHOD=ENTER gender youth_exe cardio equal
  /CONTRAST (gender)=Indicator(1)
  /CONTRAST (cardio)=Indicator(1)
  /CONTRAST (equal)=Indicator(1)
  /CRITERIA=PIN(.05) POUT(.10) ITERATE(20) CUT(.5).
```

Stata–output 4.20
To the best of my knowledge, there is no SPSS option readily available for generating marginal effects.

4.10 Chapter Exercises with Solutions

The exercises below use the data available for download on the book's website.

Exercises:

Exercise 1 (data: soccer, see appendix B of Chapter 2 for data documentation)
  1a Use match_tot as y and the following x-variables in a multiple regression: nation_dum, origin, and age. Describe your results.
  1b Add the variable pos to the model in 1a. Describe your results.
  1c Is there an interaction effect between nation_dum and age? Describe your results.
  1d Use inc_year as y and the following x-variables in a multiple regression: nation_dum and club_rank. Describe your results.
  1e Use log_inc as y and the following x-variables in a multiple regression: nation_dum, match_tot, and club_rank. Describe your results.

Exercise 2 (data: student_exercise, see Appendix C of Chapter 2 for data documentation)
  2a Use times_exer as y and the following x-variables in a multiple regression: gender, fitness_cen, sport_club, age and age-square. Describe your results.
  2b Is there an interaction effect between gender and fitness_cen? Describe your results.

Exercise 3 (data: student_exercise_motive, see Appendix A of this chapter for data documentation)
  3a Use the dummy sport_club as y and the following x-variables in a multiple regression: gender, youth_exe, and exer_easy_motive. Describe your results.
  3b Use the dummy sport_club as y and the following x-variables in a multiple *logistic* regression: gender, youth_exe, and exer_easy_motive. Describe your results.

Answers to exercises (in Stata only; see Section 4.9 for equivalent SPSS syntaxes):
Exercise 1 (data: soccer, see appendix B of Chapter 2 for data documentation)
  1a Use match_tot as y and the following x-variables in a multiple regression: nation_dum, origin, and age. Describe your results. The multiple regression is:

```
. reg match_tot i.nation_dum i.origin age
```

Source	SS	df	MS			
				Number of obs	=	240
				F(3, 236)	=	105.63
Model	845337.393	3	281779.131	Prob > F	=	0.0000
Residual	629534.403	236	2667.51866	R-squared	=	0.5732
				Adj R-squared	=	0.5677
Total	1474871.8	239	6171.0117	Root MSE	=	51.648

match_tot	Coef.	Std. Err.	t	P>|t|	[95% Conf. Interval]	
nation_dum						
yes	35.63203	7.690887	4.63	0.000	20.48047	50.78359
origin						
foreign	-41.46065	7.673479	-5.40	0.000	-56.57792	-26.34339
age	11.21185	.7808202	14.36	0.000	9.673586	12.75012
_cons	-210.77	20.09348	-10.49	0.000	-250.3555	-171.1845

Ceteris paribus throughout: Players who have appeared for their national teams have played almost 36 more matches in their careers on average than players without such experience. Foreign players have played about 41 fewer matches in their careers on average than Norwegian players. Both differences are large and suggest marked associations between the variables. A 25-year old player has played 11 more matches in his career on average than a 24-year old player. There is thus a noticeable association between the two variables. The three x-variables combined explain 57 percent of the variation (variance) in total number of matches played in the career.

1b   Add the variable pos to the model in 1a. Describe your results. The multiple regression is:

```
. reg match_tot i.nation_dum i.origin age i.pos
```

Source	SS	df	MS			
				Number of obs	=	240
				F(6, 233)	=	55.55
Model	868055.361	6	144675.894	Prob > F	=	0.0000
Residual	606816.435	233	2604.36238	R-squared	=	0.5886
				Adj R-squared	=	0.5780
Total	1474871.8	239	6171.0117	Root MSE	=	51.033

match_tot	Coef.	Std. Err.	t	P>|t|	[95% Conf. Interval]	
nation_dum						
yes	36.02876	7.657817	4.70	0.000	20.94135	51.11617
origin						
foreign	-42.41111	7.662505	-5.53	0.000	-57.50776	-27.31446
age	11.28132	.773535	14.58	0.000	9.757308	12.80534
pos						
defense	27.802	12.01627	2.31	0.022	4.127573	51.47643
midfield	34.05961	12.16511	2.80	0.006	10.09194	58.02728
attack	33.9909	12.66001	2.68	0.008	9.048187	58.93361
_cons	-241.0651	22.50931	-10.71	0.000	-285.4129	-196.7173

The coefficients or effects change little from the regression model in 1a. On average, defense players have played 28 more matches in their careers than goalkeepers *ceteris paribus*; the analogous differences between goalkeepers and midfielders and between goalkeepers and attackers are 34 and 34, respectively. There is a clear association between player position and total number of matches played in career. The increase in R^2 from 57 to 59 percent also supports this.

    1c Is there an interaction effect between nation_dum and age? Describe your results. The multiple regression is:

```
. reg match_tot i.nation_dum##c.age i.origin i.pos
```

Source	SS	df	MS		
Model	912602.208	7	130371.744	Number of obs =	240
Residual	562269.588	232	2423.57581	F(7, 232) =	53.79
				Prob > F =	0.0000
				R-squared =	0.6188
Total	1474871.8	239	6171.0117	Adj R-squared =	0.6073
				Root MSE =	49.23

match_tot	Coef.	Std. Err.	t	P>|t|	[95% Conf. Interval]	
nation_dum						
yes	-151.8088	44.43133	-3.42	0.001	-239.3492	-64.26829
age	8.858933	.9359854	9.46	0.000	7.014815	10.70305
nation_dum#						
c.age						
yes	6.807248	1.587784	4.29	0.000	3.67893	9.935566
origin						
foreign	-37.39789	7.483687	-5.00	0.000	-52.14256	-22.65321
pos						
defense	28.48688	11.59281	2.46	0.015	5.646247	51.32751
midfield	31.17347	11.75458	2.65	0.009	8.014101	54.33283
attack	30.75672	12.23597	2.51	0.013	6.648901	54.86454
_cons	-179.5553	26.02569	-6.90	0.000	-230.8322	-128.2784

The age effect is 8.86 for players who have not appeared for their national teams. For players with appearances for their national teams, the age effect is 15.67 (8.86 + 6.81 = 15.67). That is, we have a steeper regression line for the association between age and number of matches played in career among players who (also) play for their national teams. The increase in R^2 also supports this. To obtain the graph showing the interaction effect (in Stata), enter the two following commands in succession after running the regression model in question:

```
. reg inc_year i.nation_dum club_rank
```

Source	SS	df	MS		
Model	3.7026e+11	2	1.8513e+11	Number of obs =	240
Residual	7.6245e+11	237	3.2171e+09	F(2, 237) =	57.55
				Prob > F =	0.0000
				R-squared =	0.3269
Total	1.1327e+12	239	4.7394e+09	Adj R-squared =	0.3212
				Root MSE =	56720

```
------------------------------------------------------------------------------
    inc_year |      Coef.   Std. Err.      t    P>|t|     [95% Conf. Interval]
-------------+----------------------------------------------------------------
  nation_dum |
         yes |   59276.49    8056.76     7.36   0.000     43404.48      75148.5
   club_rank |   -4568.13   820.1303    -5.57   0.000    -6183.807    -2952.454
       _cons |   103742.9    8636.47    12.01   0.000     86728.83     120756.9
------------------------------------------------------------------------------
```

Ceteris paribus, we find that national team players on average earn 59,276 Euros more than non-national team players. A player for the club finishing at, say, seventh place at the end of the season earns on average 4,568 Euros less than a player for the club finishing at sixth place *ceteris paribus*. There are marked associations between both x-variables and income. The two x-variables combined explain 33 percent of the variation (variance) in yearly income.

1e   Use log_inc as y and the following x-variables in a multiple regression: nation_dum, match_tot, and club_rank. Describe your results. The command and multiple regression are:

```
. reg log_inc i.nation_dum match_tot club_rank

      Source |       SS           df       MS      Number of obs   =       240
-------------+----------------------------------   F(3, 236)       =     48.35
       Model |  50.2947926         3   16.7649309   Prob > F        =    0.0000
    Residual |  81.8266954       236   .346723286   R-squared       =    0.3807
-------------+----------------------------------   Adj R-squared   =    0.3728
       Total |  132.121488       239   .552809573   Root MSE        =    .58883


------------------------------------------------------------------------------
     log_inc |      Coef.   Std. Err.      t    P>|t|     [95% Conf. Interval]
-------------+----------------------------------------------------------------
  nation_dum |
         yes |   .5190479   .0901715     5.76   0.000      .341404     .6966918
   match_tot |   .0017734   .0005299     3.35   0.001     .0007294     .0028174
   club_rank |  -.0529426   .0085327    -6.20   0.000    -.0697527    -.0361326
       _cons |   11.20951   .0976152   114.83   0.000      11.0172     11.40181
------------------------------------------------------------------------------
```

Ceteris paribus, we find that national team players earn 68 percent more than non-national team players on average. We must apply the formula $100 \times (e^b - 1)$ to get this percentage. In the Command window in Stata, type dis 100 * (exp(0.519)-1) and press Enter to get ≈ 68.

A player having played 100 matches earns almost 2 percent more on average than a player having played 90 matches *ceteris paribus* ($0.0018 \times 10 = 0.018 \rightarrow 1.8$ percent). Finally, a player for the club finishing at, say, seventh place at the end of the season earns 5.3 percent less on average than a player for the club finishing at sixth place *ceteris paribus*. There are clear associations between all three x-variables and income. The three x-variables combined explain 38 percent of the variation (variance) in log of yearly income.

Exercise 2 (data: `student_exercise`, see appendix C of Chapter 2 for data documentation)

    2a Use `times_exer` as y and the following x-variables in a multiple regression: `gender`, `fitness_cen`, `sport_club`, `age` and age-square. Describe your results. The multiple regression is:

```
. reg times_exer i.gender##i.fitness_cen i.sport_club c.age##c.age

      Source |       SS           df       MS      Number of obs   =       644
-------------+----------------------------------   F(6, 637)       =     32.98
       Model |  415.907722         6  69.3179536   Prob > F        =    0.0000
    Residual |  1338.91021       637  2.10189986   R-squared       =    0.2370
-------------+----------------------------------   Adj R-squared   =    0.2298
       Total |  1754.81793       643  2.72911032   Root MSE        =    1.4498


------------------------------------------------------------------------------
  times_exer |      Coef.   Std. Err.      t    P>|t|     [95% Conf. Interval]
-------------+----------------------------------------------------------------
      gender |
        male |   .0521222   .1762645     0.30   0.768    -.2940075    .3982519
             |
 fitness_cen |
         yes |   1.088216   .1431224     7.60   0.000     .8071672    1.369265
             |
     gender# |
 fitness_cen |
    male#yes |   .6980461   .2395553     2.91   0.004     .2276325     1.16846
             |
  sport_club |
         yes |   .8398128   .1438636     5.84   0.000     .5573086    1.122317
         age |  -.3368372   .1104769    -3.05   0.002    -.5537802   -.1198942
             |
```

Ceteris paribus throughout: Male students exercise on average 0.41 more times per week than female students. Fitness center members exercise on average 1.33 more times per week than non–members. The analogous effect for sports club membership is 0.83 times per week. There is a U-pattern between age and number of times of exercise per week, with age = 29 years as the lowest point. We use the formula $-b_1/(2 \times b_2)$ to get this low-point for the association between age and exercising. In the Command window in Stata, type `dis (0.3507558/(2 * 0.0059923))` and press Enter to get ≈ 29.

In summary, there are clear associations between all three x-variables and times exercise per week. The three x-variables combined explain 23 percent of the variation (variance) in times of exercise per week.

    2b Is there an interaction effect between `gender` and `fitness_cen`? Describe your results. The multiple regression is:

```
. reg times_exer i.gender##i.fitness_cen i.sport_club c.age##c.age

      Source |       SS           df       MS      Number of obs   =       644
-------------+----------------------------------   F(6, 637)       =     32.98
       Model |  415.907722         6  69.3179536   Prob > F        =    0.0000
    Residual |  1338.91021       637  2.10189986   R-squared       =    0.2370
-------------+----------------------------------   Adj R-squared   =    0.2298
       Total |  1754.81793       643  2.72911032   Root MSE        =    1.4498
```

```
------------------------------------------------------------------------------
   times_exer |      Coef.   Std. Err.      t    P>|t|     [95% Conf. Interval]
--------------+---------------------------------------------------------------
       gender |
         male |   .0521222   .1762645     0.30   0.768    -.2940075    .3982519
              |
  fitness_cen |
          yes |   1.088216   .1431224     7.60   0.000     .8071672    1.369265
              |
     gender#  |
  fitness_cen |
     male#yes |   .6980461   .2395553     2.91   0.004     .2276325    1.16846
              |
   sport_club |
          yes |   .8398128   .1438636     5.84   0.000     .5573086    1.122317
          age |  -.3368372   .1104769    -3.05   0.002    -.5537802   -.1198942
              |
  c.age#c.age |   .0057211   .0019707     2.90   0.004     .0018513    .009591
              |
        _cons |   6.773726   1.480998     4.57   0.000     3.865499    9.681954
------------------------------------------------------------------------------
```

Ceteris paribus, female fitness center members exercise 1.09 more times per week than female non-members on average. *Ceteris paribus*, male fitness center members exercise 1.79 more times per week than male non-members on average (1.09 + 0.70 = 1.79). The difference between fitness center members and non-members in terms of times exercising per week differ for the genders; we have an interaction effect. Finally, there is practically speaking no gender difference in times of exercise for non-members of fitness centers (0.052).

Exercise 3 (data: student_exercise_motive, see appendix A of this chapter for data documentation)

3a   Use the dummy sport_club as *y* and the following *x*-variables in a multiple regression: gender, youth_exe, and exer_easy_motive. Describe your results. The multiple regression is:

```
. reg sport_club i.gender youth_exe exer_easy_motive

      Source |       SS           df       MS      Number of obs   =       510
-------------+----------------------------------   F(3, 506)       =     33.28
       Model | 18.3356546          3  6.11188486   Prob > F        =    0.0000
    Residual | 92.9270905        506  .183650377   R-squared       =    0.1648
-------------+----------------------------------   Adj R-squared   =    0.1598
       Total | 111.262745        509  .218590855   Root MSE        =    .42854

------------------------------------------------------------------------------
   sport_club |      Coef.   Std. Err.      t    P>|t|     [95% Conf. Interval]
--------------+---------------------------------------------------------------
       gender |
         male |   .1272694   .0401826     3.17   0.002     .0483241    .2062147
    youth_exe |   .0516238   .0066688     7.74   0.000     .0385218    .0647257
 exer_easy_m~e|   .0474192   .0158038     3.00   0.003      .01637    .0784685
        _cons |  -.2303037   .0639908    -3.60   0.000    -.3560239   -.1045834
------------------------------------------------------------------------------
```

Ceteris paribus throughout: Male students on average have a 12.7 percentage points greater probability of being a sports club member than female students. A student with the score of, say, five on the youth sports variable has on average a 5.2 percentage points greater probability of being a sports club member than a student with the score of four. A student scoring, say, four on the exercise motive has on average a 4.7 percentage points greater probability of being a sports club member than a student scoring three.

   3b   Use the dummy sport_club as y and the following x-variables in a multiple
        logistic regression: gender, youth_exe, and exer_easy_motive. Describe
        your results. The multiple logistic regression and its marginal effects become:

```
. logit sport_club i.gender youth_exe exer_easy_motive

Output omitted

Logistic regression                               Number of obs    =        510
                                                  LR chi2(3)       =      96.81
                                                  Prob > chi2      =     0.0000
Log likelihood = -271.89757                       Pseudo R2        =     0.1511

------------------------------------------------------------------------------
  sport_club |      Coef.   Std. Err.      z    P>|z|     [95% Conf. Interval]
-------------+----------------------------------------------------------------
      gender |
        male |   .6833809   .2139075     3.19   0.001      .26413    1.102632
   youth_exe |   .3243538   .0466864     6.95   0.000     .2328501    .4158575
exer_easy_m~e |   .2599929    .089528     2.90   0.004     .0845212    .4354646
       _cons |   -4.27745    .477646    -8.96   0.000    -5.213619   -3.341281
------------------------------------------------------------------------------

. margins, dydx(*)

Average marginal effects                          Number of obs    =        510
Model VCE    : OIM

Expression   : Pr(sport_club), predict()
dy/dx w.r.t. : 1.gender youth_exe exer_easy_motive

------------------------------------------------------------------------------
             |            Delta-method
             |     dy/dx   Std. Err.      z    P>|z|     [95% Conf. Interval]
-------------+----------------------------------------------------------------
      gender |
        male |   .1275926   .0402781     3.17   0.002     .048649    .2065363
   youth_exe |   .0583862   .0070164     8.32   0.000    .0446342    .0721381
exer_easy_m~e |   .0468007   .0156053     3.00   0.003    .0162149    .0773866
------------------------------------------------------------------------------
Note: dy/dx for factor levels is the discrete change from the base level.
```

Ceteris paribus throughout: Male students on average have a 12.8 percentage points greater probability of being a sports club member than female students. A student with the score of, say, five on the youth sports variable has on average a 5.8 percentage points greater probability of being a sports club member than a student with the score of four. A student scoring, say, four on the exercise motive has on average a 4.7 percentage points greater probability of being a sports club member than a student scoring three. The multiple logistic regression and its linear counterpart yield very similar answers in a qualitative sense.

Notes

1 If we randomly distribute two different types of questionnaires, imitating an RCT, we call it a survey experiment. See Section 4.6 for more on this.
2 This kind of question/statement goes by the name of a Likert scale.
3 Remember that the bivariate regression coefficient is the average change in y given a one-unit increase in x. In the multiple regression case, we simply add the holding constant clause to this interpretation.
4 The negative constant -1.02 is nothing to worry about. The constant has a literal interpretation only when zero is a legitimate value for all x-variables in the regression.
5 Of course, the regression coefficients for exercise motivation (x_2) and exercise importance (x_3) variables might also change in magnitude when a fourth and a fifth x-variable enter the regression.
6 Metaphorically once more: We may think of the statistics program first running a quality-price regression for beers having an alcohol level of 5 percent. Next, the statistics program runs a second and similar quality-price regression for beers having an alcohol level of 6 percent and so on for the remaining alcohol levels in the data. The average of all these regression coefficients is -0.172.
7 The *ceteris paribus* clause is a convenient shorthand to avoid spelling out the names of all the other x-variables included in the regression model.
8 Strictly speaking, R^2 refers to the variance in y, but this goes for the same in practice.
9 The adjusted R^2 adjusts for the fact that more x-variables in a model almost by definition increases R^2.
10 The regression line has a U-shape and a bottom-point when the coefficient for b_2 has a positive sign. Nothing changes with more x-variables in the regression model except for the entering of the *ceteris paribus* clause.
11 R^2-comparison between the two models is one criterion. Another criterion is that the coefficient for the square term (i.e., x^2) should be statistically significant, which it is. We dive into statistical significance in Chapter 5.
12 R^2 for the linear model is 0.202, and R^2 for the quadratic model is 0.259 (not shown).
13 In some research traditions, such as the determinants of income, the log regression model is in fact the default model rather than the plain vanilla regression model.
14 We just add the *ceteris paribus* clause in the usual manner in the multiple regression case.
15 Some researchers, especially economists, often log x to make both x and y measured in logs. When associating $\log x$ and $\log y$, the interpretation is the average change in y in percent given a 1-percent increase in x. We call this elasticity.
16 I focus on only x_1 and b_1 for presentational ease. The same reasoning applies in the multiple regression case for x_2 and b_2 and so on.
17 I assume that a linear regression model is ok in the following for ease of presentation. That said, the same principle applies to non-linear settings.
18 Kennedy (2002) has another name for Occam's razor: the KISS principle: Keep It Sensibly Simple! Yet we might want to examine any possibility of non-parallel regression lines from an exploratory point of view, as in the case for linearity versus non-linearity.
19 Some prefer moderator variable or moderator term; I personally prefer interaction variable/term.
20 I explain the commands for interactions and non-linearity in more depth in Section 4.9.
21 The third and final strike is that the regression coefficient for the interaction term should be statistically significant, which it is. More on statistical significance in Chapter 5, Section 5.7.
22 Entering more x-variables does not complicate matters for the two examples of interaction effects in this section. I only omit such variables for presentational ease.
23 Kahneman (2011) is the obvious source for all matters on the two authors' collaborations. You also find the experiment I present here in Kahneman's book, although I use the modified version developed by Gelman and Nolan (2017).
24 The instructor prepares two versions of the assignment, say 50 with "65" and 50 with "10" and shuffles them randomly into a pile of 100 assignments before distributing them (again randomly) in class. In other words, the random number generator is bogus!
25 More precisely, as we shall see in Section 5.7, it is evidence against the idea that a nonsense anchor has *no* effect on the subsequent assessment.

26 If randomization is not successful, however, control variables may help as for observational data. There is one more justification for using control variables in experiments: to increase the precision of the treatment variable's effect. We find this in the reduction of the treatment variable's standard error from 4.128 to 4.073. This reduction is trivial in the present case, however.

27 I focus on dummy y-variables. It is also possible to do regression on categorical y-variables with more than two unordered outcomes, such as yes, no, and do not know. See the further reading section for this.

28 Some might protest here, claiming that we should do the so-called logistic regression when y is a dummy. I will address this claim later in the section.

29 Linear regression between a dummy y and a dummy x always reproduces a similar cross-tabulation between the two variables. This no longer holds when more x-variables enter the regression.

30 I think Angrist and Pischke's (2009) influential book on econometrics marks the revival of the LPM model. By the way, econometrics is just the economists' more hotshot name for statistical data analysis.

31 See Thrane (2020) for a more in-depth treatment of logistic regression analysis.

32 If 140 out of the 644 students (i.e., about 22 percent) are members of a sports club, the odds of being such a member is 0.28 $(0.22/(1 - 0.22) = 0.282)$.

33 Stata may also report odds-ratios; cf. Section 4.9.

34 SPSS has to the best of my knowledge no option to produce marginal effects based on logistic regression.

35 We generally do not know the correct result in real life. That is, we do not know if the true gender difference is 17.6 or 16.9 percentage points. If we did, there would be no reason for trying to estimate the gender difference in the first place!

Appendix A
Student Exercise Motive Data

Data documentation for the data `student_exercise_motive`; a survey questionnaire data from a random sample of students attending a Norwegian university college in 2018. Variable names are in **bold** typeface. N = 510.

`fitness_cen`

Member of fitness center: no = 0, yes = 1

`sport_club`

Member of sport club: no = 0, yes = 1

`health`

Your physical health condition: ok = 0, good = 1, very good = 2

`qol`

Your quality of life: ok = 0, good = 1, very good = 2

`youth_exe`

In your youth before you started studying, to what extent were you involved in sports that required a lot of physical exercising (to a very small extent = 1, to a very great extent = 10)?

`exer_imp_qol`

Statement/question: Exercising three times per week is important for my quality of life! Answer scale: totally disagree = 1, disagree = 2, neither/nor = 3, agree = 4, totally agree = 5

`exer_easy_motive`

Statement/question: I easily find the motivation to exercise! Answer scale: totally disagree = 1, disagree = 2, neither/nor = 3, agree = 4, totally agree = 5

`age`

Age in years

gender

Gender: female = 0, male = 1

snuff

Snuffing (moist snuff): never = 0, sometimes/daily = 1

hours_exer

Hours of weekly exercise (in hours)

Appendix B
Student Tourism Data

Data documentation for the data stud_tourism; a survey questionnaire data for a random sample of Norwegian students on their summer vacation trip of longest duration. Variable names are in **bold** typeface. N = 444.

los

Length of trip in days (duration)

book_time

Number of weeks from booking to start of trip

accom

Type of accommodation on trip: Commercial = 0, Private = 1

trav_party

Travel party on trip: Alone = 0, Friends = 1, Partner = 2, Other = 3

pay_meth

Main payment method on trip: Cash = 0, Card = 1

destin

Trip destination: Nordic country = 0, Beyond Nordic country = 1

type_trip

Type of trip: Non-package trip = 0, Package trip = 1

tot_spend

Total personal expenditures on trip in Euros

Appendix C

Red Wine Data

Documentation for the data `red_wine`; a data set of red wines quality-tested by a Norwegian newspaper some years ago. Variable names are in **bold** typeface. N = 218.

quality

Taste quality of the wine on a scale from 60 (tasteless) to 100 (perfect taste): The Parker Scale

district

Wine district: Burgundy = 1, Bordeaux = 2, Languedoc/Roussillon = 3, Rhone = 4, Other = 5

age

The wine's age, that is, the number of years it has been stored

in_store

Wine available in store: No, must be ordered = 0, Yes = 1

price

Price per bottle of red wine in Euros

log_price

Natural logarithm of **price**

5 Inferential Research Questions

5.1 Introduction and Chapter Overview

Chapters 2–4 concerned only what happened inside our data – as expressed perhaps a bit cryptically at the end of Sections 2.1, 3.1, and 4.1. This chapter spins around one key question: How can we be sure what we find inside our data is something that also takes place outside our data? This is an inferential research question. To answer it properly, I take a rather circumstantial route with many stops along the way. Why bother with this? The short answer is that it needs to be done!

Sections 5.2–5.5 build the foundation for the main event of this chapter, namely Section 5.7 on hypothesis testing and the assessment of statistical significance. Section 5.6, in contrast, is an intermediate section verifying that the many claims of Sections 5.2 to 5.5 are correct. Some do not quarrel with statements like 'statisticians have proved that …' and accept this without further ado. Others, among them myself, find such statements borderline infuriating. I have thus written Section 5.6 for those who do not accept matters at face value and partly for myself!

The presentation of the topics in this chapter, and particularly those in Section 5.7, takes an implicit notice of the by now long-lasting 'statistical significance testing controversy,' for lack of a better term. Yet I postpone this important topic for an explicit and separate section, namely Section 5.8.

Section 5.9 summarizes the chapter and lists the key learning points in a familiar manner, whereas Section 5.10 is the usual coverage of do-file and syntax-file commands. Section 5.11 provides some exercises with solutions.

Note to instructors! I have written this chapter with one particular type of student in mind: One who finds math difficult and encounters inferential statistics for the first time. This chapter is thus not written for instructors who already know the material! I emphasize this point more strongly than before because it has a direct bearing on the pedagogical order in which I present the various topics and on the level of the preciseness in their explications. Other instructors, likely those more skilled in statistics than me, might claim I over-simplify. My first response to such a claim is, 'Yes, I do.' My second response is that over-simplifying is often necessary to get the main message through. A third response, someone might say, is that over-simplifying with success oftentimes is very difficult to do.

5.2 Samples, Populations, and Random Sampling

Two of the first concepts you encounter when picking up any book on statistics are 'samples' and "populations." In this book thus far, however, I have not discussed these terms. Why? The answer is that they complicate matters unnecessarily and that we do

DOI: 10.4324/9781003252559-5

not need them to get a firm grip on what practical statistical analysis is mainly about. In this respect, we have already been through the basics without even bothering to define the two concepts properly. At present, however, it is essential to get some formalities in place as a foundation for what comes next.

At the beginning of Chapters 2–4, I claimed that my upcoming comments regarding the statistical results pertained only to what happened inside our data. There were two reasons for this: First, I wanted a focus on what happens inside our data, and a focus on what might happen (or not) outside our data takes that focus away. Second, we always analyze only our data. We oftentimes have a research ambition stretching beyond our particular data, but such an inference always involves a leap (of faith) from what we actually know about our data to what we do not know about what happens in the world beyond our data.

The inference from what we know (our data) to what we may reasonably speculate about (the world outside our data) is crucial. The main problem is that we in research are seldom interested in what happens in our data per se.[1] On the contrary, our research interests most often pertain to the world outside our data. Yet since it is impossible to examine this outside world directly, we use our data to do so indirectly. The success or failure of this strategy rests on the foundation that our data resemble, or are representative of, the outside world. This question of representativeness is critical. If our data are unrepresentative of the outside world, it follows that they help us little in terms of making valid inferences about this world. If we predict the height of an adult male based on the average height of the 20 adult women in our data, our prediction will be too low because women on average are shorter than men. That is, our inference will be off since our data and the world beyond our data do not match up. This begs the question of how to make data representative of the outside world. Please welcome the distinction between a sample and a population and the idea of random sampling.

I am very sure you have heard the terms "sample" and "population" before. Indeed, what I have called data thus far is often a sample. For example, our student exercise data is a sample of 644 students. The world beyond our data is in this case the student population, as in all the students attending the university in question. We may thus deduce that a population is all the units of a particular thing sharing at least one common trait – such as enrollment in the same university. A sample, in contrast, is some smaller proportion or fraction of such a population – as in 644 out of all the university students in our case.

Populations often refer to many people with a common characteristic: all adults in Australia, all adult women in Canada, and all basketball players in the NBA. Yet a population does not have to contain people in a technical sense. All transactions on the Danish stock exchange in 2020 are a population, and so are all car commutes by Canadian female drivers in 2020. Furthermore, all hockey games played in the 2020/21 season of the NHL are a population. Any smaller proportions of these populations are by definition a sample.

If a sample was always our data and the world beyond our data was always a population, I could simply have claimed that we in quantitative research use a sample to make an inference about an unknown population. Yet this is not necessarily the case in present-day statistical analysis.[2] Consider, for example, our data on Norwegian soccer players that include all the players in the top-tier league. In this case, there is no need for a sample-to-population inference because the data by definition are a (known) population.[3] Still, these population data may by themselves serve as a relevant source of data for answering more general research questions, that is, inferential research questions stretching beyond our particular population data.

Inferential statistics has much to do with the quantification of uncertainty. One issue is important at the outset in this regard: As long as we focus on what is going on inside our data – which could be a sample, a population, or whatever – questions about uncertainty are typically irrelevant. That is, statistics programs do the calculations for the data correctly. Uncertainty, in other words, comes up when we wish to make an inference from our data to the world beyond our data. Furthermore, in the case of a sample versus a population, such an inference is probabilistic in certain circumstances.[4] We look at these certain circumstances in more detail below.

We prefer a representative sample to an unrepresentative sample for obvious reasons. The big-ticket question is thus how to make a sample representative of a larger population. The best strategy is to draw a *random* sample. Random sampling means that every unit in the population has a known and roughly similar chance of ending up in the sample. Suppose 16,500 students attend the university from which we collected our student exercise data. Suppose also that each student had a student number and that we used a random function to select 644 numbers without replacement. (Any computer can make a random draw from a list of 16,500 numbers.) The first student thus had a 1/16,500 chance of being selected into the sample. For the second student the chance was 1/16,499, for the third it was 1/16,498, and so on for the remaining 641 students. In short, the 16,500 students had a known and approximately similar chance of ending up in the sample receiving the invitation to take part in the survey.[5]

The random sampling procedure is the mechanism that makes a sample representative of a larger population rather than just *a* sample. In a manner of speaking, random sampling creates a miniature model of the population of interest.[6] We then use this miniature model (i.e., the sample) to make an inference about the full-scale model (i.e., the population) under certain, probabilistic conditions. These conditions need some clarification, though, and this is the topic of the two next sections.

5.3 Repeated Sampling and the Normal Distribution

5.3.1 Repeated Sampling

Andrew, a friend of mine, is an avid fan of Norwegian top-tier soccer. He also has a systematic streak. Before the 2017 soccer season, he chose 25 matches to attend live at the stadiums. Moreover, he chose his 25 matches randomly from a list of all the 240 matches to be played in the 2017 season, that is, the population. He also gathered information on the matches he attended, such as the number of goals scored by the home team, the number of goals scored by the away team, the number of spectators etc. By the end of 2017, Andrew had thus compiled information on a set of variables for 25 of the matches in the Norwegian top-tier league of 2017. Furthermore, his data were a random sample of the population of soccer matches played in the 2017 season of the top-tier Norwegian soccer league.

Goal-scoring is one of the features in soccer Andrew finds interesting. Based on the scoring records for both teams, he created the variable total number of goals scored per match (goals). The name of the data file is Andrew_data. The summary descriptive statistics for this goal-scoring variable appear in Stata-output 5.1. The analogous SPSS-output looks very similar, tells the exact same story, and is thus redundant. This is also the case for most of the other Stata-outputs in this chapter. In the few instances where the Stata-outputs and the SPSS-outputs differ, however, I will also present the analogous SPSS-outputs.

```
. sum goals if season == 0
```

Variable	Obs	Mean	Std. Dev.	Min	Max
goals	25	2.84	1.624808	1	7

Stata-output 5.1 Descriptive statistics for the total number of goals per match in Andrew's random sample of the 2017 season.

We note that the two teams playing against each other scored 2.84 goals combined per match on average. The range for this goal-scoring variable is one to seven goals. The SD is 1.62. There is absolutely nothing new here compared with what we have been through so far in the book, except perhaps for the fact that the unit of analysis – that is, the soccer match – comprises two teams.

Suppose the above story is true, but now for an alternate reality experiment. Imagine Andrew making his computer do the random draw of the 25 matches for the 2017 season but forgetting to write down the list of 25 numbers (matches) because he suddenly remembered being late to a meeting. So next time at home, he makes a second random draw of 25 matches from the list of 240 matches and gets 25 new numbers. (This is sampling with replacement.) Some of the new numbers might have been on Andrew's original list, but the majority will be new numbers. Now, picture Andrew attending these 25 alternate matches and compiling the "same" data as before. What would the mean, minimum, and maximum be for the total goals variable this time? Because this is an alternate reality scenario, we do not know. Yet a good guess is *close to the results of the first and actual draw* in Stata-output 5.1. We would not anticipate the results to be identical, however, because of the expected random variation between Andrew's actual sample and our imagined sample.

Let us step up our alternate reality experiment a notch. The second time Andrew makes the random draw the fire alarm goes off. He shuts down his computer in a hurry without saving and runs out of the apartment. Back again (it was a false alarm!), Andrew makes a third random draw. Now picture Andrew attending this *second* alternate list of 25 matches and compiling the data in the usual way. What would the mean and so on be for the total goals variable this time? Again, we cannot know for sure. Yet a good guess is once more "*close to results of the first and actual draw* in Stata-output 5.1." Again, however, we would not anticipate the results to be identical because of the expected random variation between Andrew's actual sample and our two imagined samples.

You get the picture of what happens in a fourth random draw, so I do not have to come up with any more bad excuses for Andrew. The results will typically be close to the results of similar statistical analyses based on other random samples from the same population. This thought experiment of making consecutive random draws from the same population and doing the same statistical analysis – say, to find a mean – for the various samples one obtains is fundamental in statistics. We call it *repeated sampling*.

Think back on our 644 students in the exercise data. Imagine making a second random draw to get a new random sample of 644 students. What would be the results if we were to repeat the analyses in this book on this alternate random sample? We cannot know for sure, but a good guess in line with the reasoning above is that the results in the main would be close to the ones we got for the sample we actually analyzed. And so on for a third, fourth, and fifth alternate random sample etc. Yet we would not expect the results for the various samples to be identical because of random variation between samples.

Let us not forget about Andrew yet. Above I described the procedure for getting his random sample for the 2017 season. However, he did a similar thing for the 2018, 2019, and 2020 seasons. His total data are thus a random sample of 100 matches (25 × 4 seasons) from the population of 960 soccer matches in 2017–2020 (240 × 4 seasons). Summary descriptive statistics for the variable total goal-scoring per match appear in Stata–output 5.2.

```
. sum goals

    Variable |        Obs        Mean    Std. Dev.        Min        Max
-------------+-------------------------------------------------------------
       goals |        100        2.96    1.607558          0          7
```

Stata-output 5.2 Descriptive statistics for the total number of goals per match in Andrew's random sample of the 2017–2020 seasons.

We note that the two teams scored 2.96 goals combined per match on average, with a range from zero to seven goals. The SD is close to 1.61. Both results are very close to the results for the 2017 season in Stata–output 5.1, which should come as no surprise by now.

What is the average of total goals per match in the population of 960 matches? The short answer is that we do not know. The longer one is that we do not know this because there would be no reason for taking a sample to begin with if we did! That said, we could speculate. A good guess in this regard – the best guess, even – is somewhere in the proximity of Andrew's sample mean of 2.96. I show why this is a good guess below.

5.3.2 The Normal Distribution

Repeated sampling is one bedrock of statistics. Another bedrock is the *normal distribution*. We met the normal distribution in Section 2.6 when scrutinizing the distribution for the alcohol level variable in the beer data. Owing to statisticians, we know one fundamental fact about a normally distributed continuous variable in a large sample: 95 percent of its units lie in the interval mean ± two SDs.[7] We can thus make the following rough inference for the alcohol level variable in Stata-output 5.3, assuming for convenience that it has a normal distribution and pertains to a large sample: 95 percent of the Christmas beers in the data have an alcohol level in the range between 5.06 and 11.32 percent (8.187 ± 2 × 1.566 = 8.187 ± 3.132 = [5.055, 11.319]).[8]

```
. sum alch_perc

    Variable |        Obs        Mean    Std. Dev.        Min        Max
-------------+-------------------------------------------------------------
   alch_perc |         75    8.186667    1.565622          5         12
```

Stata-output 5.3 Descriptive statistics for alcohol level in the Christmas beer data.

In practice, we often relate the normal distribution to a *standardized* variable. A standardized variable has a mean of zero by definition, and one SD is called z.[9] Figure 5.1 depicts the normal distribution for a standardized variable and its relationship to z (or z-score). We see that 95 percent of the distribution for a standardized and normally distributed variable lies in the interval $0 \pm 2\ z$.

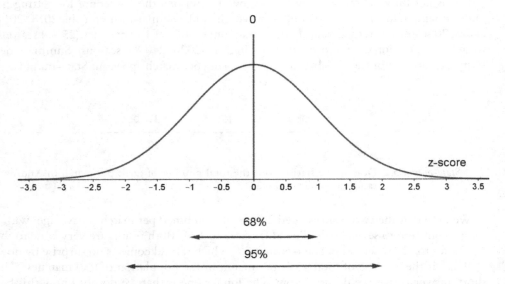

Figure 5.1 Percentage of units under the normal curve for a standardized variable; 95% of the units lies in the interval $0 \pm 2\ z$.

5.3.3 Combining Repeated Sampling and the Normal Distribution

How does the idea of repeated sampling relate to the normal distribution? We return to Andrew's interest in goal-scoring to answer this, but we address the sample collection from another angle. We know the population contains exactly 960 matches. It is thus doable to compile a list of 960 numbers – one for each match – and to make a random draw of 100 numbers from this list. Suppose we did just that, found the match statistics for the 100 matches on the Internet (unlike Andrew), and calculated the mean of the total goals variable. Assume this mean was 3.05; a result different from Andrew's but not by much. Suppose we made a second random draw of 100 matches and repeated the process for the second sample. Assume the second result for the total goals variable was 2.93. Third random draw, third sample, third data compilation, third analysis, and a third mean for total goals: 2.88. Once more: fourth random sample ... and a fourth mean for the total goals variable: 2.90. Table 5.1 displays the results of this imagined repeated sampling project stopping at 100 random samples.

Now, think of Table 5.1 as a data set in which "Mean of total goals" is a variable and each sample is a unit. The variable "Mean of total goals" thus has a mean and an SD. Furthermore, and thanks to statisticians once more, we know *a priori* that this variable has a normal distribution: If we actually were to do this repeated sampling project and calculate the mean for each new sample, the distribution of means would follow the normal distribution in Figure 5.1.[10] (We return to this normal distribution in Section 5.6.) Based on this knowledge, we may calculate an interval containing the true and unknown mean of the total goals variable in the population. Please welcome the 95 percent CI!

Table 5.1 Results of an imagined repeated random sampling project.

Random samples:	N =	Mean of total goals
Andrew's sample	100	2.96
Our first sample	100	3.05
Our second sample	100	2.93
Our third sample	100	2.88
Our fourth sample	100	2.90
...	100	2.84
...	100	2.92
... Our sample 99	100	3.02
... Our sample 100	100	2.98

5.4 The 95 Percent CI for Descriptive Statistics: Means and Proportions

5.4.1 Means

If the "data" in Table 5.1 were real, we could calculate the mean and the SD to find the interval containing 95 percent of the matches with respect to the mean of the total goals variable: the mean ± two SDs.[11] There is only one problem with this approach; the data are not real! For this reason, we have no mean or SD to base our calculations on. Yet the statisticians save the day once more with their knowledge. Thanks to the random sampling principle, we know that Andrew's sample most often is representative of the unknown population. We may therefore use his sample mean of 2.96 as an *estimate* for the total goals variable's unknown population mean. One problem solved; we have the mean we need! The next problem is the SD. We mentioned in Chapter 2 that the SD measured the size of the variation around the mean of a continuous variable. We could thus use Andrew's sample-SD for total goals as an estimate for this variable's unknown population-SD – just as did for the mean. Yet this is inefficient because there is much more variation in the total number of goals variable from match to match than there is variation in the *mean* of the total number of goals variable from repeated sample to repeated sample.[12] However, we may compute an estimate of the latter repeated-sample SD based partly on Andrew's sample-SD. The name of this repeated-sample SD is the standard error (SE). The formula to obtain the SE of a mean is

$$\frac{s}{\sqrt{n}},$$

where s is the SD of the total goals variable in Andrew's sample in this case, and n is the number of matches in Andrew's sample in this case. From Stata–output 5.2 the SE thus becomes

$$= \frac{1.6076}{\sqrt{100}} = 0.16076 \approx 0.161$$

To find the interval containing 95 percent of the matches with respect to the mean of the total goals variable in the population, we apply the formula mean ± 2 SEs. That is, we substitute the SD with the SE. This interval becomes

$$2.96 \pm 2 \times 0.161 = 2.96 \pm 0.322 \approx [2.64, \ 3.28].$$

The approximate interpretation is that we are 95 percent confident the interval from 2.64 to 3.28 goals contains the true and unknown mean of the total number of goals variable in the population. We call this the 95 percent confidence interval (CI). More precisely, in repeated samples of the same size from the same population, the true population mean will lie in the interval from 2.64 to 3.28 goals 95 percent of the time. I will mostly use the former rough interpretation for convenience in the examples to come.

Stata-output 5.4 verifies our calculations. The output displays the SE (Std. Err.) and the 95 percent CI on the right. Stata's 95 percent CI is based on more decimals than my calculation and on multiplication with a number slightly smaller than 2.0, that is, 1.984. SPSS-output 5.1 reports approximately the same interval in addition to some of the measures we covered in Chapter 2.

```
. mean goals

Mean estimation                    Number of obs    =        100

             |      Mean    Std. Err.    [95% Conf. Interval]
-------------+------------------------------------------------
       goals |      2.96    .1607558     2.641026    3.278974
```

Stata-output 5.4 The 95 percent CI for total number of goals per match in the 2017–2020 seasons.

Explore

Case Processing Summary

	Cases					
	Valid		Missing		Total	
	N	Percent	N	Percent	N	Percent
goals	100	100,0%	0	0,0%	100	100,0%

Descriptives

			Statistic	Std. Error
goals	Mean		2,96	,161
	95% Confidence Interval for Mean	Lower Bound	2,64	
		Upper Bound	3,28	
	5% Trimmed Mean		2,93	
	Median		3,00	
	Variance		2,584	
	Std. Deviation		1,608	
	Minimum		0	
	Maximum		7	
	Range		7	
	Interquartile Range		2	
	Skewness		,364	,241
	Kurtosis		-,180	,478

SPSS-output 5.1 The 95 percent CI for total number of goals per match in the 2017–2020 seasons.

We have analyzed the variable students' total hours of exercise per week on several occasions. What is this variable's 95 percent CI? The upper part of Stata-output 5.5 contains the numbers we need to calculate by hand; the lower part makes the calculation for us. With 95 percent confidence, the unknown mean of total hours of weekly exercise lies in the range between 4.54 and 5.02 hours in the population. (SPSS shows the exact same results of course.)

```
. sum hours_exer

    Variable |        Obs        Mean    Std. Dev.         Min         Max
-------------+--------------------------------------------------------------
  hours_exer |        644     4.78028    3.147652           0          16

. mean hours_exer

Mean estimation                        Number of obs     =        644

             |        Mean   Std. Err.      [95% Conf. Interval]
-------------+--------------------------------------------------------------
  hours_exer |     4.78028   .1240349      4.536717    5.023842
-------------+--------------------------------------------------------------
```

Stata-output 5.5 Descriptive statistics and the 95 percent CI for hours of weekly exercise.

5.4.2 *Proportions*

The examples thus far were CIs for means. We use the same approach to find the 95 percent CI of a proportion, but we use another formula to find its SE. The formula to obtain the SE of a proportion is

$$\sqrt{\frac{p(1-p)}{n}},$$

where p is the sample proportion in question and n is the number of units. Suppose we want to find the SE for the proportion of sports club members in the student exercise data. The sample numbers of interest are 0.2174 (i.e., 21.74 percent are members, cf. Stata output 4.16) and 644 (i.e., the number of students). The SE thus becomes

$$\sqrt{\frac{0.2174(1-0.2174)}{644}} = 0.01625$$

We now have the information we need to compute the 95 percent CI, namely,

$$2.96 \pm 2 \times 0.161 = 2.96 \pm 0.322 \approx [2.64, 3.28].$$

We are 95 percent confident the interval from 18.5 to 25 percent contains the true and unknown proportion of sports club members in the population. Stata again uses more decimals and multiplies the SE with a number slightly less than 2.0 (i.e., 1.96) to obtain the 95 percent CI; cf. Stata-output 5.6. (The analogous SPSS-output shows the same.)

```
. prop sport_club

Proportion estimation                    Number of obs    =       644

----------------------------------------------------------------
                 |                                  Logit
                 | Proportion   Std. Err.    [95% Conf. Interval]
-----------------+----------------------------------------------
     sport_club  |
            no   |  .7826087    .0162536     .7490092    .8128342
           yes   |  .2173913    .0162536     .1871658    .2509908
----------------------------------------------------------------
```

Stata-output 5.6 The 95 percent CI for the proportion of students being members of sports clubs.

The three previous examples – two for a mean and one for a proportion – have one thing in common: They tell us we know less precise things about a population than we do about a random sample from such a population. If this is true in general (which it is!), it must also hold for the difference between two means, the difference between two proportions, and for variable associations in general. This is coming up next.

5.5 The 95 Percent CI for Variable Associations

5.5.1 Difference in Two Means

Stata–output 5.7 presents the gender-specific means in total exercise hours per week for the student data. We note the familiar and average gender difference of 1.447 hours per week (5.7149 − 4.2680 = 1.4469), with males being more active than females.

```
. oneway hours_exer gender, t

                |        Summary of hours_exer
        gender  |       Mean    Std. Dev.        Freq.
----------------+------------------------------------
        female  |  4.2680288   2.7558764          416
          male  |  5.7149123   3.5807053          228
----------------+------------------------------------
         Total  |  4.7802795   3.147652           644

                     Analysis of Variance
      Source            SS          df       MS            F       Prob > F
----------------------------------------------------------------------------
Between groups       308.325468      1    308.325468      32.65     0.0000
 Within groups       6062.33408    642    9.44288798
----------------------------------------------------------------------------
      Total          6370.65955    643    9.90771314

Bartlett's test for equal variances:  chi2(1) =  20.9168   Prob>chi2 = 0.000
```

Stata-output 5.7 Hours of weekly exercise by gender. One-way ANOVA.

What is the 95 percent CI for the mean gender difference in the population? We proceed along the same lines as before to answer this question. First, we find the SE of the mean difference. The formula is

$$\sqrt{\frac{s_1^2}{n_1} + \frac{s_2^2}{n_2}},$$

where, in this case, s_1 is the SD for females, n_1 is the number of females, s_2 is the SD for males, and n_2 is the number of males. From Stata-output 5.7 the SE thus becomes

$$\sqrt{\frac{2.756^2}{416} + \frac{3.581^2}{228}} = 0.2729 \approx 0.273.$$

From this, we obtain the 95 percent CI

$$1.447 \pm 2 \times 0.273 = 1.447 \pm 0.546 \approx [0.90, 1.99].$$

We are 95 percent confident the interval from almost one to almost two hours contains the true and unknown mean gender difference in exercise hours in the population. Note that the CI, in this case, is always larger than zero; it does not include a zerohour mean difference between the genders. We may tentatively deduce that the male students in the population exercise on average more hours per week than female students. (We return to this conclusion more formally in Section 5.7.)

Again, Stata uses more decimals and a number less than 2.0 (i.e., 1.96) to calculate the analogous results, which appear in Stata-output 5.8. The diff row at the bottom of the output presents the relevant numbers. The negative sign reflects that Stata in this case subtracts the male mean from the female mean. SPSS-output 5.2 presents the same results in an analogous manner in the bottom row.

```
. ttest hours_exer, by(gender) une

Two-sample t test with unequal variances
-----------------------------------------------------------------------------
   Group |     Obs        Mean    Std. Err.   Std. Dev.   [95% Conf. Interval]
---------+-------------------------------------------------------------------
  female |     416    4.268029     .135118    2.755876    4.002428    4.53363
    male |     228    5.714912     .237138    3.580705    5.247639    6.182185
---------+-------------------------------------------------------------------
combined |     644     4.78028    .1240349    3.147652    4.536717    5.023842
---------+-------------------------------------------------------------------
    diff |              -1.446883    .2729309              -1.983543    -.910224
-----------------------------------------------------------------------------
    diff = mean(female) - mean(male)                          t =   -5.3013
Ho: diff = 0                      Satterthwaite's degrees of freedom =  376.607

    Ha: diff < 0                  Ha: diff != 0                   Ha: diff > 0
 Pr(T < t) = 0.0000        Pr(|T| > |t|) = 0.0000          Pr(T > t) = 1.0000
```

Stata-output 5.8 The 95 percent CI for the mean gender difference in weekly exercise hours.

T-Test

Group Statistics

	gender	N	Mean	Std. Deviation	Std. Error Mean
hours_exer	female	416	4,2680	2,75588	,13512
	male	228	5,7149	3,58071	,23714

Independent Samples Test

		Levene's Test for Equality of Variances		t-test for Equality of Means					95% Confidence Interval of the Difference	
		F	Sig.	t	df	Sig. (2-tailed)	Mean Difference	Std. Error Difference	Lower	Upper
hours_exer	Equal variances assumed	26,663	,000	-5,714	642	,000	-1,44688	,25321	-1,94410	-,94966
	Equal variances not assumed			-5,301	376,607	,000	-1,44688	,27293	-1,98354	-,91022

SPSS-output 5.2 The 95 percent CI for the mean gender difference in weekly exercise hours.

5.5.2 Difference in Two Proportions

We find the 95 percent CI for a difference between two proportions along the same route as we did for the difference between two means, but we use a different formula to calculate the SE (cf. Agresti, 2018). Yet since we typically prefer to carry out a comparison of proportions by another statistical technique, to be described in Section 5.7, I will not get into this. Differences in means and proportions are two types of variable associations; a regression coefficient is a third. We close the present section by looking at this.

5.5.3 A Regression Coefficient

What is the 95 percent CI for a regression coefficient? To answer this, we first find the coefficient's SE and then proceed in the usual way: the coefficient $\pm$ 2 $\times$ SEs. Yet since calculating the SE of a regression coefficient is a complicated affair, we leave this to the statistics program and proceed directly to the association between gender and exercise hours in Stata-output 5.8. Stata-output 5.9 presents the analogous analysis regression-style. We note that male students on average exercise 1.447 hours more per week than female students – as we already know.

```
. reg hours_exer i.gender

      Source |        SS           df       MS          Number of obs   =        644
-------------+----------------------------------        F(1, 642)       =      32.65
       Model |   308.325468          1   308.325468      Prob > F        =     0.0000
    Residual |   6062.33408        642   9.44288798      R-squared       =     0.0484
-------------+----------------------------------        Adj R-squared   =     0.0469
       Total |   6370.65955        643   9.90771314      Root MSE        =     3.0729


  hours_exer |      Coef.   Std. Err.      t    P>|t|     [95% Conf. Interval]
-------------+----------------------------------------------------------------
      gender |
        male |   1.446883   .2532102     5.71   0.000     .9496631    1.944104
       _cons |   4.268029   .1506627    28.33   0.000     3.972178    4.563888
------------------------------------------------------------------------------
```

Stata-output 5.9 Hours of weekly exercise by gender in the student exercise data. Linear regression.

The SE (Std. Err.) of the regression coefficient is 0.2532, and the 95 percent CI on the right goes from 0.95 to 1.94. The reason why the regression-SE is slightly different from the ANOVA-SE in Stata-outputs 5.7 and 5.8 lies in the mathematical properties of regression. We make the same general inference as before, however: Because the CI does not include zero (i.e., gender equality), we tentatively conclude that male students appear to exercise more hours per week on average than female students in the population. SPSS-output 5.3 reports the same conclusion, but as usual in a somewhat different format.

Model Summary

Model	R	R Square	Adjusted R Square	Std. Error of the Estimate
1	,220[a]	,048	,047	3,07293

a. Predictors: (Constant), gender

ANOVA[a]

Model		Sum of Squares	df	Mean Square	F	Sig.
1	Regression	308,325	1	308,325	32,652	,000[b]
	Residual	6062,334	642	9,443		
	Total	6370,660	643			

a. Dependent Variable: hours_exer
b. Predictors: (Constant), gender

Coefficients[a]

Model		Unstandardized Coefficients B	Std. Error	Standardized Coefficients Beta	t	Sig.	95,0% Confidence Interval for B Lower Bound	Upper Bound
1	(Constant)	4,268	,151		28,328	,000	3,972	4,564
	gender	1,447	,253	,220	5,714	,000	,950	1,944

a. Dependent Variable: hours_exer

SPSS-output 5.3 Hours of weekly exercise by gender in the student exercise data. Linear regression.

5.6 Verifying That Random Sampling and the Central Limit Theorem Work as Promised

A random sample is a means to an end; a population is that end. That is, we study a known random sample to get more knowledge about an unknown population. The previous sections have taken us up the steps in a bottom–up approach from sample to population based on the ideas of random sampling, the normal distribution, and theoretical and imagined statistical reasoning. Now it is time to show how this actually works by doing the exact opposite, that is, a top–down population-to-sample verification exercise. To be clear, the normal situation is that we do not know what happens in the population of interest, and this is why we draw a random sample to begin with. In the upcoming and particular case, however, we have access to data for the entire population. We may thus retrace our steps in the bottom–up process by going in the opposite top–down direction.

Table 5.2 Results of an actual repeated random sampling project from a population of 960 soccer matches. $N = 100$ for each sample; 50 samples with replacement.

Repeated samples	Mean of total goals	Repeated samples	Mean of total goals
Andrew's sample	2.96	Sample 26	3.04
Sample 2	3.15	Sample 27	2.77
Sample 3	2.90	Sample 28	3.14
Sample 4	2.71	Sample 29	2.92
Sample 5	2.72	Sample 30	2.91
Sample 6	2.52	Sample 31	2.98
Sample 7	2.64	Sample 32	3.15
Sample 8	3.17	Sample 33	3.00
Sample 9	2.96	Sample 34	3.13
Sample 10	2.99	Sample 35	3.20
Sample 11	2.65	Sample 36	3.00
Sample 12	2.70	Sample 37	3.00
Sample 13	2.77	Sample 38	3.29
Sample 14	2.86	Sample 39	2.92
Sample 15	2.85	Sample 40	2.67
Sample 16	3.05	Sample 41	2.61
Sample 17	2.66	Sample 42	3.18
Sample 18	2.81	Sample 43	3.05
Sample 19	2.85	Sample 44	2.88
Sample 20	3.06	Sample 45	3.09
Sample 21	3.06	Sample 46	2.91
Sample 22	2.91	Sample 47	2.90
Sample 23	3.18	Sample 48	2.74
Sample 24	3.27	Sample 49	2.75
Sample 25	3.08	Sample 50	3.10

You might have figured out that "Andrew" is no friend of mine; that he is a fictitious character. You might have had a hunch I made him up for this Chapter. I did. Furthermore, you might have guessed that I have a data set covering the entire population of soccer matches in the 2017–2020 period ($N = 960$). I have.[13] It is thus easy to imagine how I used a random draw function to select "Andrew's sample" of 100 matches, and how I computed the mean of the total goals variable. I can do this just as easily 49 more times – which I actually have done. That is, I turned the imagined repeated sampling project mentioned in Section 5.3 into an actual repeated sampling project. Table 5.2 presents the results in this regard. (The table uses four columns instead of two to save space.) On top and left, we find Andrew's result for the total goals variable: the mean of 2.96. For the second random sample, the analogous mean became 3.15. The mean for the third random sample became 2.90 and so on for the remaining 46 samples.

Think of Table 5.2 as a data set having only one continuous variable – "Mean of total goals" – and 50 units. That is, one unit equals one (repeated) sample. Scanning through the data, we see that no sample mean is smaller than 2.52 (sample 6) or larger than 3.29 (sample 38); the range is thus 0.77. Section 5.3 said the sampling distribution has a normal distribution. Figure 5.2 verifies this using a Kernel density plot for the 50 mean values in Table 5.2. We note the symmetrical and bell-shaped curve roughly corresponding with the superimposed normal distribution. The Kernel plot for the 50 mean values would have been even smoother and more perfectly aligned with the normal distribution if we had taken 100 repeated samples rather than just 50.

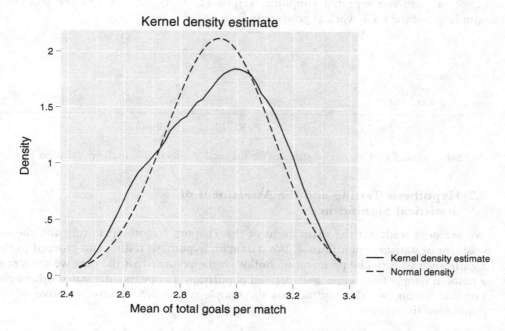

Figure 5.2 Kernel density plot for the variable mean of total goals in Stata–output 5.10, based on the data in Table 5.2.

Stata-output 5.10 displays the mean of the 50 mean values in Table 5.2. The output shows that this repeated-sample mean is 2.94 goals, with a repeated-sample SD of 0.188. The analogous numbers were 2.96 and 1.608 for Andrew's sample; cf. Stata-output 5.2. As expected, the two means are close. Furthermore, and again as expected, there is much more variation in total goal-scoring from match to match in Andrew's sample (1.61) than there is variation in the mean of total goal-scoring among repeated samples (0.188). Interestingly, we note that the estimated SE based on Andrew's sample – 0.161; cf. Stata-output 5.4 – is very close to the repeated-sample SD of 0.188.

```
. sum totgoal_mean

    Variable |        Obs        Mean    Std. Dev.         Min         Max
-------------+--------------------------------------------------------------
totgoal_mean |         50      2.9362    .1879198        2.52        3.29
```

Stata-output 5.10 Descriptive statistics for the variable mean of total goals (totgoal_mean) based on the data in Table 5.2. The name of the data file is repeat _ samp _ 50.

What about the population mean? Stata-output 5.11 shows that the population mean for the total goals variable is 2.93. This is not exactly the same as in Andrew's sample

(2.96) or as in our repeated sampling exercise (2.94). But it is very close. Random sampling and the CLT work as promised!

```
. sum goals

    Variable |        Obs        Mean    Std. Dev.        Min        Max
-------------+-------------------------------------------------------------
       goals |        960    2.927083    1.694897          0          9
```

Stata-output 5.11 Descriptive statistics for the total goals variable in the population.

5.7 Hypothesis Testing and the Assessment of Statistical Significance

We are now ready for the main event of this chapter: hypothesis testing and the assessment of statistical significance. We start with hypothesis testing and proceed to the significance aspect. The presentation builds on the premise that the data we analyze is a random sample from some well-defined population if not explicitly stated otherwise. For this reason, we also assume that the sample we analyze is representative of the population in question.

5.7.1 Hypothesis Testing: Preliminaries

Hypothesis testing concerns at least two variables, and this x-and-y case is our point of departure.[14] Specifically, we believe x affects y for some plausible reason.[15] We may for example hypothesize that male students exercise more hours per week than female students because of a stronger (male) exercise motivation. In this case, x is gender, y is hours of exercise, and motivation is the reason. In a related spirit: Students active in sports in their youth exercise more hours per week than students less active in sports in their youth because of habit formation. The x is youth exercise involvement, y is hours of exercise, and the reason is habit formation.[16] That said, any hypothesis always refers to a population. The hypotheses above should thus have included an "in the population" phrase I omitted for the sake of readability. Yet although a hypothesis always refers to a population, we test it against what we find in a sample from this population.[17] Before going into the actual testing, however, we must first make a detour concerning the two types of hypotheses we face in research: the alternative hypothesis and the null hypothesis.

5.7.2 The Alternative Hypothesis and the Null Hypothesis

The alternative hypothesis refers to what we believe in regarding the association between x and y in the population *prior* to the statistical analysis of this association for the sample. We denote this alternative hypothesis H_1. For the gender and exercising case, we thus have

H_1: Male students exercise more hours per week than female students.[18]

The null hypothesis specifies what we do not believe in for the population, or the "opposite" of what we believe in, namely

H_0: Male students exercise for the same number of hours per week as female students.

The alternative hypothesis generally suggests some form of non-zero association between x and y in the population, whereas the null hypothesis implies a zero association between x and y. The key point is that we always test the null hypothesis![19] We test the null hypothesis for the population against what we find in a random sample from this population. If what we find in the random sample differs from what we should find in the population according to the null hypothesis (i.e., a zero association between x and y), we reject the null hypothesis and get indirect support for the alternative hypothesis. That is, we get support for a non-zero association between x and y in the population in keeping with what we believe in *a priori*, that is, the alternative hypothesis.

A small and vital word in the above paragraph is "differs." How do we assess if what we find in the sample differs from what we should find in the population according to the null hypothesis? The answer is that we use tests of statistical significance for exactly this purpose.

5.7.3 *Statistical Significance: Preliminaries*

When examining the association between x and y in a random sample, we know that any association we find will not be of exactly the same magnitude as the equivalent population association. Suppose the sample association in question is a regression coefficient of 0.5. In the population, the analogous coefficient could be, say, 0.2, 0.3, 0.4, 0.6, 0.7, or 0.8 merely because of random differences between the sample and the population. Could the population coefficient also be zero, implying that x has no effect on y in the population? A statistical significance test tries to answer this question, and I will now show this for the association between gender and exercise hours.

5.7.4 *Gender and Hours of Exercise Once More: ANOVA*

The null hypothesis suggests that male students exercise for the same number of hours per week as female students in the population. We saw in Stata-outputs 5.7 and 5.8 that male students on average exercised 1.45 hours more per week than female students. (The 95 percent CI went from 0.90 to 1.99.) To find out if random differences between the population and the sample may account for this 1.45-hour mean gender difference, we turn the 95 percent CI on its head. That is, we compute a 95 percent CI for the null hypothesis. The formula to get this 95 percent CI for the null hypothesis, which goes by the name of the 95 percent *acceptance region* for the null hypothesis, is

$0 \pm 2 \times$ SEs.

Since we already know from the calculation based on Stata-output 5.7 that the SE in question is 0.273, we thus obtain the 95 percent acceptance region

$$0 \pm 2 \times 0.273 = 0 \pm 0.546 \approx [-0.546, \ 0.546].$$

Given a correct null hypothesis in the population – that is, gender equality in the mean of total exercise hours – we are 95 percent confident the interval from −0.55 to 0.55 hours contains the mean gender difference in exercise hours in the population. The 1.45-hour mean gender difference in our sample is outside of this acceptance region – just as the 95 percent CI for the mean gender difference did not include zero in Section 5.5. We, therefore, reject the null hypothesis and get indirect support for the alternative hypothesis suggesting that male students on average exercise more hours per week than female students in the population. The association between gender and exercise hours is *statistically significant.* Alternatively (and a bit simplified), we might say the observed mean gender difference in exercise hours appears not to be the result of random differences between the sample and the population.

5.7.5 Gender and Hours of Exercise Once More: Regression Analysis

We do a similar thing in the regression case. The null hypothesis – that is, gender equality in the mean of total exercise hours in the population – implies that the regression coefficient for the gender dummy variable should be zero. The 95 percent acceptance region for this null hypothesis is

$$0 \pm 2 \times \ SEs.$$

We already know that the SE in question is 0.253; cf. Stata-output 5.9 or SPSS-output 5.3. We thus get the 95 percent acceptance region

$$0 \pm 2 \times 0.253 = 0 \pm 0.506 \approx [-0.506, \ 0.506].$$

Given the null hypothesis implying a gender regression coefficient of zero in the population, we are 95 percent confident the interval from −0.51 to 0.51 hours contains the gender regression coefficient in the population. Our gender coefficient of 1.45 hours for the sample is outside of this acceptance region. Again, we reject the null hypothesis and get indirect support for the alternative hypothesis. The gender regression coefficient has a statistically significant effect on hours of exercise, implying that male students on average exercise more hours per week than female students in the population.

The two prior examples compared an acceptance region for a null hypothesis in a population with a statistical result for a sample. Because the sample result was outside of the acceptance region for H_0, we rejected it and got indirect support for H_1.[20] The association was thus statistically significant. But suppose the sample result was inside of the acceptance region of H_0, we would have then kept H_0 and gotten no support for H_1. The association between x and y would then have been statistically *insignificant* – or not statistically significant. Thankfully, finding out if a mean difference between two groups or a regression coefficient, or any other statistical association, is statistically significant, is often more of a mechanical action in research than making the calculations above. Let us, therefore, look at this for our familiar examples.

5.7.6 *From t-values to p-values*

To find out if a variable association is statistically significant, we typically divide the expression for the association in question by its SE. This provides us with a *t*-value. We return to the regression between gender and hours of weekly exercise, which reappears in Stata–output 5.12. By dividing the regression coefficient (1.44883) by its SE (0.2532102), we get the *t*-value of 5.71 reported in column t; see also SPSS-output 5.3.

```
. reg hours_exer i.gender

      Source |       SS           df       MS            Number of obs   =        644
-------------+----------------------------------         F(1, 642)       =      32.65
       Model |  308.325468          1  308.325468        Prob > F        =     0.0000
    Residual |  6062.33408        642  9.44288798        R-squared       =     0.0484
-------------+----------------------------------         Adj R-squared   =     0.0469
       Total |  6370.65955        643  9.90771314        Root MSE        =     3.0729

------------------------------------------------------------------------------
 hours_exer |      Coef.   Std. Err.      t    P>|t|     [95% Conf. Interval]
-------------+----------------------------------------------------------------
      gender |
        male |   1.446883   .2532102     5.71   0.000     .9496631    1.944104
       _cons |   4.268029   .1506627    28.33   0.000     3.972178     4.56388
------------------------------------------------------------------------------
```

Stata-output 5.12 Hours of weekly exercise by gender in the student exercise data. Linear regression.

The *t*-value is important in inferential statistics involving means – such as for the regression between gender and hours of exercise.[21] In large samples, *t*-values roughly correspond to the *z*-values we saw in Figure 5.1.[22] This has one very important implication: A *t*-value larger than ± 2 for a regression coefficient (or an ANOVA mean difference) implies outside of the 95 percent acceptance region for the null hypothesis. That is, a *t*-value larger than ± 2 suggests a statistically significant association between *x* and *y*. In contrast, a *t*-value in the range from −2 to 2 suggests that such an association is not statistically significant. The *t*-value for our gender regression coefficient is 5.71. This is outside of the 95 percent acceptance region for the null hypothesis, as shown in the upper part of Figure 5.3. The association is hence statistically significant.

The 95 percent CI for a mean, for a difference between two means, for a regression coefficient, and for the null hypothesis' acceptance region have one thing in common: the insistence on making the correct inference 95 percent of the time in repeated sampling. In practice, however, we focus on the flip side of this 95 percent certainty level. That is, we are dead set on *not* making the wrong inference more than 5 percent of the time. We refer to this threshold value as the *5 percent significance level*. This 5 percent significance level, in turn, corresponds with a *t*-value of ± 2.[23] How does this relate to the *t*-value of 5.71 of our gender coefficient? The bottom of Figure 5.3 illustrates. We note that the *t*-value lies way outside of the acceptance region for the 5 percent significance level coinciding with a *t*-value in the region from –2 to 2.

Figure 5.3 also shows the acceptance region for the 1 percent significance level, that is, a *t*-value in the range from –2.58 to 2.58. We may think of the 1 percent significance level as raising the bar even higher in terms of not making the wrong inference. Our gender regression coefficient's *t*-value of 5.71 is on the outside of the 1 percent significance level as well.

The idea of not making the wrong call more than five times out of a 100 suggests we may say something more exact about this probability of wrongdoing. Welcome to the *p*-value, where *p* is short for probability. The *p*-value is the flip side of the *t*-value. A large *t* yields a low *p*, and a small *t* yields a large *p*. For this reason, since we most often do research in the hope of finding a statistically significant association between *x* and *y*, we hope to find large *t*-values/small *p*-values. (If for some particular reason we do a statistical analysis in which the alternative hypothesis is that *x* does *not* affect *y*, we thus hope for a small *t*/a large *p*.)

All statistics programs report *p*-values. We find the *p*-value on the right of the *t*-value under the column heading P>|t| in Stata-output 5.12. The *p*-value for the gender regression coefficient is 0.000, which implies less than 0.0001 or less than 0.01 percent.[24] In other words, the probability of obtaining a gender regression coefficient of 1.445 (or a larger one) is below 0.01 percent if the true coefficient in the population is zero. Alternatively, given no mean gender difference in the population (i.e., H_0), it is very unlikely to obtain a mean gender difference of 1.445 hours in a random sample from this population. We analogously find the *p*-value in SPSS-output 5.3 on the right of the *t*-value under the Sig. column heading. The result in SPSS is of course similar to the result in Stata.

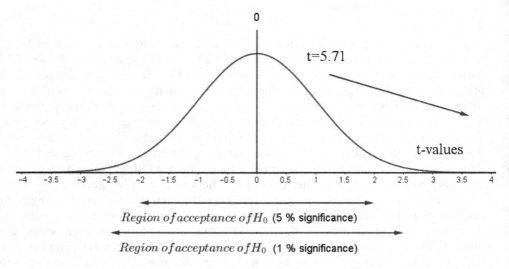

Figure 5.3 Region of acceptance for the null hypothesis and its relationship with *t*-values.

The *p*-value plays an important role in statistics, but it is often misinterpreted. The strictly correct interpretation in our case is the probability of finding a gender regression coefficient of 1.445, or a larger one, if the true coefficient in the population is zero.[25] It is important to note what this *p*-value is not: It is not the probability of random chance creating the statistical association in the sample – which many seem to believe. The *p*-value measures something less relevant, namely, "… in a world where your hypothesis isn't true, how likely is it that pure noise would give you results like the ones you have, or ones with an even larger effect?" (Ritchie, 2020, p. 88).[26] That said, if the *p*-value has a very low probability, it makes sense believing that something else more systematic is going on – such as the stated reason for the alternative hypothesis to begin with.

Stata–output 5.13 presents a regression between hours of exercise and the student status variable. The constant (or reference) is the single students, and the coefficient for the boyfriend/girlfriend group suggests that this group exercises 0.09 hours less per week than the single students. The difference between the cohabiting/married group and the single students is 0.57 hours in the same negative direction. Yet neither of these coefficients have *p*-values below 0.05. We conclude there is no statistically significant association between exercise hours and the student status variable in the population.[27] That said, the coefficient for the cohabiting group is not far from being significant at the 5 percent level, with a *p*-value of 0.10 or 10 percent. (SPSS of course yields similar results.)

```
. reg hours_exer i.status

      Source |       SS           df       MS         Number of obs   =       644
-------------+----------------------------------        F(2, 641)       =      1.37
       Model |  27.1549227         2   13.5774614       Prob > F        =    0.2543
    Residual |  6343.50463       641   9.89626307       R-squared       =    0.0043
-------------+----------------------------------        Adj R-squared   =    0.0012
       Total |  6370.65955       643   9.90771314       Root MSE        =    3.1458


  hours_exer |      Coef.   Std. Err.      t    P>|t|     [95% Conf. Interval]
-------------+----------------------------------------------------------------
      status |
       bf/gf |   -.094188   .2794994     -0.34   0.736    -.6430332    .4546571
     cohabit |  -.5724324   .3476654     -1.65   0.100    -1.255133    .1102683
             |
       _cons |   4.907295   .1734354     28.29   0.000     4.566725    5.247865
```

Stata-output 5.13 Hours of exercise by student status in the student exercise data. Linear regression.

5.7.7 Significance Tests in Multiple Regression

The multiple regression in Stata–output 5.13 has two dummy variables: the boyfriend/girlfriend group versus the single group and the cohabiting group versus the single group. For this reason, we have one alternative hypothesis and one null hypothesis for each of the two dummy *x*-variables in the regression model. Save for this, nothing is new in the multiple *x*-variable case compared with the bivariate case.

5.7.8 The Margin of Error

The 95 percent CI and the 95 percent acceptance region for the null hypothesis have one common feature: the expression $\pm 2 \times$ SEs. This is the *margin of error*. Looking at the formulas for the SEs in Sections 5.4 and 5.5, we note they have sample size in their denominators. (This applies to all SE-formulas.) This suggests that the SE and thus the error margin decreases by necessity as sample size increases – all else being equal. Smaller error margins thus means narrower or more precise CIs and acceptance regions. To double the precision, as in reducing the error of margin by 50 percent, we need to increase the sample size fourfold. This has practical implications we return to in Section 5.8.

5.7.9 One-Sided and Two-Sided Significance Tests

So-called two-sided or two-tailed tests of significance dominate statistical research. A two-sided test means that we reject the null hypothesis if the t-value is larger than ± 2; cf. Figure 5.3. Sometimes, however, we use one-sided tests of significance. Below I clarify when to adopt the two-sided and when to adopt the one-sided test for the multiple regression in Stata-output 5.13. We may have two different alternative hypotheses for the difference in exercise hours between, say, the single and the cohabiting students:

H_{1a}: Cohabiting students exercise *fewer* hours per week than single students.
H_{1b}: Cohabiting students and single students exercise for an *unequal* number of hours per week.

H_{1a} is a directional hypothesis, whereas H_{1b} is non-directional. A directional hypothesis is thus a more specific expectation regarding the association between x and y. A directional H_1 entails a one-sided test of significance. This implies that we reject H_0 only when the t-value is *either* larger than 2 *or* larger than −2, depending on the direction (i.e., plus or minus) of H_1. For a non-directional H_1, in contrast, we reject H_0 whenever the t-value is larger than ± 2. Suppose H_{1b} is at stake. The p-value in this regard, as we already know, is 0.10. But suppose H_{1a} was what mattered. Since we now reject H_0 only when the t-value is larger than −2 on the left side of Figure 5.3, we should divide the p-value in the statistics program by two: $0.10/2 = 0.05$ or 5 percent. Thus, given H_{1a}, the difference in exercise hours between cohabiting and single students is statistically significant at the 5 percent level exactly.[28] The upshot is that it is "easier" to get a significant result for a directional H_1 than it is for a non-directional H_1. The state of knowledge concerning the research question of interest determines if a directional hypothesis is more prudent than a non-directional one to begin with.

5.7.10 t-Values and p-Values in Small Samples

The CLT tells us that the larger the random sample, the more it resembles the population from which it was drawn. This has one key implication: Random differences between populations and samples are more pronounced for smaller samples than for larger samples – all else being equal. In order not to make an error more than five times out of a 100 when rejecting a null hypothesis for a small sample, we must therefore guard ourselves against these enlarged random differences. This defense entails raising the bar as in demanding a higher *t*-value to reject the null hypothesis. We already know that the *t*-value of 2, or 1.96 to be exact, refers to large samples. Some critical *t*-values for various sample sizes are:

Sample size:[29]	5 percent significance (two-sided):	1 percent significance (two-sided):
N = 30	2.042	2.750
N = 60	1.980	2.660
N = 100	1.984	2.626
N = 120	1.980	2.617
Large sample	1.960	2.576

Suppose we do a bivariate regression analysis for a sample of about 30 units. In this case, we reject the null hypothesis for a 5 percent significance level when the *t*-value for the regression coefficient exceeds ± 2.042 given a non-directional alternative hypothesis.

5.7.11 Using Asterisks (*) to Denote Statistical Significance Level

It has become almost mandatory to use asterisks to denote statistical significance levels throughout the world of quantitative research. Although many are skeptical of this practice (more on the critical aspects of significance testing in Section 5.8), it would still be a severe oversight not to mention this practice. The three levels of statistical significance typically used are:

One asterisk:	*	Significant at the 5 percent level
Two asterisks:	**	Significant at the 1 percent level
Three asterisks	***	Significant at the 0.1 percent level

Some researchers and research fields use a 7 percent or 10 percent significance level as the lowest form of certainty for rejecting a null hypothesis when analyzing small samples. One asterisk (*) typically equals the 7-percent or the 10-percent level in such cases.

5.7.12 p-Values for Cross-Tabulations

The procedures for calculating *p*-values for regressions and ANOVAS are similar, which explains why I have put the latter in the notes. The cross-table is another story. We looked at the association between gender and the propensity of being a sports club member using a cross-tabulation in Section 3.2; cf. Stata output 3.1 or SPSS–output 3.1. This analysis reappears in Stata–output 5.14, with a small text-extension below the table.

```
. tab sport_club gender, col chi2


             |        gender
 sport_club  |    female       male  |       Total
-------------+----------------------+----------
         no  |       356        148  |         504
             |     85.58      64.91  |       78.26
-------------+----------------------+----------
        yes  |        60         80  |         140
             |     14.42      35.09  |       21.74
-------------+----------------------+----------
      Total  |       416        228  |         644
             |    100.00     100.00  |      100.00


         Pearson chi2(1) =   36.9667   Pr = 0.000
```

Stata-output 5.14 Sports club membership by gender in the student exercise data. Cross-tabulation, with chi-square test. N = 644.

The null hypothesis states that male and female students are sports club members to the same extent in the population. That is, the null hypothesis implies a zero gender difference. If this H_0 is correct in the population, 22 percent of the females and 22 percent of the males in the sample *should* be sports club members – in sync with the total column on the right of the table. Yet we find that 35 percent of the male students in the sample are sports club members, whereas the analogous percentage for the female students is 14 percent. The significance test below the table asks and answers if random chance differences between the sample and the population might explain this 21 percentage points gender difference given a correct H_0. The *p*-value on the right (Pr = 0.000) is less than 0.0001 or less than 0.01 percent. The random-chance explanation is very unlikely. We reject the null hypothesis and get indirect support for the alternative hypothesis suggesting some form of gender difference. SPSS reports similar information under the cross-table in SPSS-output 5.4. The *p*-value is displayed under the Asymptotic Significance column heading.

The significance tests in Stata-output 5.14 and SPSS-output 5.4 is a Pearson chi-square test. The chi-square value is almost 37 in our 2 × 2 cross-table case. How do we obtain this? The numbers without decimals in Stata-output 5.14 are the actual or observed number of students in each cell. Stata-output 5.15 also includes the *expected* number of students in each cell (in italics) given a correct null hypothesis of gender equality in the population. (The numbers are of course similar in SPSS.)

The chi-square value of approximately 37 is the difference (or deviation) between the observed (O) and the expected (E) frequencies according to the formula

$$\sum \frac{(O-E)^2}{E},$$

where *O* and *E* refer to the frequencies in each cell.

sport_club * gender Crosstabulation

			gender		
			female	male	Total
sport_club	no	Count	356	148	504
		% within gender	85,6%	64,9%	78,3%
	yes	Count	60	80	140
		% within gender	14,4%	35,1%	21,7%
Total		Count	416	228	644
		% within gender	100,0%	100,0%	100,0%

Chi-Square Tests

	Value	df	Asymptotic Significance (2-sided)	Exact Sig. (2-sided)	Exact Sig. (1-sided)
Pearson Chi-Square	36,967[a]	1	,000		
Continuity Correction[b]	35,762	1	,000		
Likelihood Ratio	35,639	1	,000		
Fisher's Exact Test				,000	,000
Linear-by-Linear Association	36,909	1	,000		
N of Valid Cases	644				

a. 0 cells (0,0%) have expected count less than 5. The minimum expected count is 49,57.

b. Computed only for a 2x2 table

SPSS-output 5.4 Sports club membership by gender in the student exercise data. Cross-tabulation with chi-square test. $N = 644$.

```
. tab sport_club gender, col chi2 exp


           |       gender
sport_club |    female       male |     Total
-----------+----------------------+----------
        no |       356        148 |       504
           |     325.6      178.4 |     504.0
           |     85.58      64.91 |     78.26
-----------+----------------------+----------
       yes |        60         80 |       140
           |      90.4       49.6 |     140.0
           |     14.42      35.09 |     21.74
-----------+----------------------+----------
     Total |       416        228 |       644
           |     416.0      228.0 |     644.0
           |    100.00     100.00 |    100.00

          Pearson chi2(1) =  36.9667   Pr = 0.000
```

Stata-output 5.15 Sports club membership by gender in the student exercise data. Cross-tabulation, with chi-square test and expected frequencies in italics. N = 644.

The chi-square test is the appropriate significance test for a cross-tabulation in the same way a *t*-test is the appropriate test for the individual *x*-variable in regressions and ANOVAs. A special feature of the chi-square test, however, is that we always should take the cross-table's number of cells into account when examining the chi-square value. The reason is that larger cross-tables "automatically" get larger chi-square values than smaller tables because the formula summarizes the deviations in each cell. To account for the size of a cross-table, we consider its degrees of freedom. The smallest cross-table, that is, the 2 × 2 table, has one degree of freedom.[30] For this 2 × 2 table the chi-square value must be larger than 3.84 to reject the null hypothesis at the 5-percent significance level (two-sided). We thus needed a chi-square of at least 3.84 to reject our null hypothesis, whereas 37 was what we got. Some other critical values for larger cross-tables appear below. Note, however, that cross-tables larger than, say, 12 cells are hard to read and most often require a very large sample size.

Number of cells:	Degrees of freedom:	5 percent significance (two-sided):
Six cells	2	5.991
Eight cells	3	7.815
Nine cells	4	9.488
Ten cells	5	11.07
Twelve cells	6	12.59

Suppose for the student population we have an alternative hypothesis suggesting gender inequality in terms being a member of a fitness center. The null hypothesis is thus gender parity. Stata-output 5.16 tests this H_0.

The probability of being a fitness center member is almost the same for male and female students: 53 versus 50 percent. The chi-square is much lower than 3.84 (i.e., 0.57),

```
. tab fitness_cen gender, col chi2

fitness_ce |         gender
        n |     female       male |      Total
-----------+----------------------+----------
       no |        210        108 |        318
          |      50.48      47.37 |      49.38
-----------+----------------------+----------
      yes |        206        120 |        326
          |      49.52      52.63 |      50.62
-----------+----------------------+----------
    Total |        416        228 |        644
          |     100.00     100.00 |     100.00

          Pearson chi2(1) =    0.5707   Pr = 0.450
```

Stata-output 5.16 Fitness center membership by gender in the student exercise data. Cross-tabulation, with chi-square test. N = 644.

and the p-value is thus much larger than 0.05 (i.e., 0.45). We dare not reject the null hypothesis of gender parity in the population. Random chance might very well have generated this tiny gender difference assuming a correct H_0.

5.8 Critical Aspects of Significance Testing

Section 5.1 gave away a clue. There has been much discussion of statistical significance testing in recent years. In particular, many have criticized how researchers routinely carry out significance testing and its assessment within the Null Hypothesis Significance Testing (NHST) framework.[31] In addition, bad practice has not gotten any better because very many books on statistics, as well as journal papers, actually misunderstand what is going on in a significance test and what we may learn – or not learn – from such a test. Now is not the time nor the place for a full account of the critique of the NHST approach in modern-day research, but I will bring up some of its main topics relevant for the newbie statistical analyst.

5.8.1 *The Null Hypothesis Is Often Stupid!*

The null hypothesis most often comes in one of three guises: no effect, no association, or no group difference. Oftentimes this is naïve if not plain stupid. Suppose you have read ten studies on gender differences in exercise hours among older adults in the age range of 20 to 80 years. Suppose further that the mean gender difference in these studies (combined) was 1.1 hours in men's favor. Imagine this being the state of knowledge before doing our regression between gender and exercise hours. Does it make sense to compare the 1.45-hour mean gender difference in our sample to a zero gender difference as per the null hypothesis? My answer is no. Moreover, when we have reason to believe a correct null hypothesis is something other than zero in the population, we should if possible test it against this more plausible non-zero value. Stata-output 5.17 presents the familiar regression again, followed by a significance test in which the null hypothesis is a 1.1-hour mean gender difference.

The p-value 0.171 says we cannot reject the null hypothesis implying an average 1.1-hour gender difference in the population:[32] We cannot claim that our mean gender difference in exercise hours in the student population is larger than the analogous mean gender difference among older adults. But suppose the mean gender difference in exercise hours among older adults was 0.9 hours. Testing against this 0.9-hour gender difference, we obtain the p-value 0.031 (result not shown). That is, against a H_0 suggesting a 0.9-hour gender difference, we reject the null and get indirect support for the alternative hypothesis that the mean gender difference in exercise hours in the student population is larger than the analogous mean gender difference among older adults. The key point is this: Whenever we have the opportunity to test an idea against a sharper null hypothesis, we should grab it rather than testing it against a nonsensical zero-effect hypothesis in the NHST framework.

```
. reg hours_exer i.gender

      Source |       SS           df       MS              Number of obs   =        64
-------------+------------------------------              F(1, 642)       =      32.6
       Model |  308.325468          1  308.325468         Prob > F        =    0.0000
    Residual |  6062.33408        642  9.44288798         R-squared       =    0.048
-------------+------------------------------              Adj R-squared   =    0.046
       Total |  6370.65955        643  9.90771314         Root MSE        =    3.072


  hours_exer |      Coef.   Std. Err.      t    P>|t|     [95% Conf. Interval
-------------+----------------------------------------------------------------
      gender |
        male |   1.446883   .2532102     5.71   0.000     .9496631    1.94410
       _cons |   4.268029   .1506627    28.33   0.000     3.972178    4.56388
-------------------------------------------------------------------------------

. test i1.gender = 1.1

 ( 1)  1.gender = 1.1

       F(  1,    642) =      1.88
            Prob > F =    0.1712
```

Stata-output 5.17 Hours of weekly exercise by gender in the student exercise data. Linear regression followed by a significance test for the null hypothesis that the gender coefficient equals 1.1 hours.

5.8.2 Statistical Significance Is a Function of Sample Size: "Everything" Becomes Significant in Large Samples!

We saw in Section 5.7 that the margin of error – that is, ± 2 × SEs – was smaller in large samples because the SE decreases with increasing sample size. To illustrate the flip side of this phenomenon, I drew a 25 percent random sample from the student exercise data and "repeated" the gender and exercise hours regression from Stata-output 5.17. The new results appear in Stata-output 5.18 and are of course similar in SPSS.

```
. reg hours_exer i.gender

      Source |       SS           df       MS              Number of obs   =        16
-------------+------------------------------              F(1, 159)       =       9.3
       Model |  73.7906832          1  73.7906832         Prob > F        =    0.002
    Residual |  1252.41429        159  7.87681941         R-squared       =    0.055
-------------+------------------------------              Adj R-squared   =    0.049
       Total |  1326.20497        160  8.28878106         Root MSE        =    2.806


  hours_exer |      Coef.   Std. Err.      t    P>|t|     [95% Conf. Interval
-------------+----------------------------------------------------------------
      gender |
        male |   1.421429    .464408     3.06   0.003     .5042245    2.33863
       _cons |   4.328571   .2738929    15.80   0.000     3.787634    4.86950
-------------------------------------------------------------------------------
```

Stata-output 5.18 Hours of weekly exercise by gender in a 25 percent random sample from the student exercise data. Linear regression.

We note a very similar gender coefficient: 1.421. In contrast, the gender coefficient's SE is 0.464 or almost twice as large as in the actual sample of 644 students (i.e., 0.253). Since we obtain the t-value by dividing the regression coefficient by its SE ($1.421/0.464 \approx 3.06$), it follows that a smaller SE makes it "easier" to get a t-value larger than ± 2 given a fixed value for the regression coefficient.[33] That is, a regression coefficient of a fixed size has an easier time getting statistically significant for a large sample compared with a small sample. Everything, so to speak, becomes statistically significant if the sample is large enough! For a small sample, in contrast, a regression coefficient must be quite large to get statistically significant due to the enlarged SE. This point also relates to the next point.

5.8.3 Statistically Significant Effect ≠ Important Effect

A significant association or difference does not mean a strong association or a large difference. The p-value refers to a null hypothesis implying a zero association or difference in a population; it says nothing about the strength of the association between x and y in a sample. That is, a significant association does not imply a strong association or a large difference in substantive terms. Nor does $p = 0.001$ imply a stronger association or a larger difference than $p = 0.04$. This also relates to the next point.

5.8.4 The 5 Percent Significance Level Is Arbitrary

Someone once said God loves 0.051 as much as he loves 0.050, although this is tough to verify. This statement points to the arbitrary threshold for what we call "statistically significant" and what we call "not statistically significant" or "insignificant" in research. Combined with the fact that the p-value largely is a function of sample size, such arbitrariness is even more questionable. To avoid yes/no thinking in statistical significance assessments, which is much too common, many researchers recommend evaluating statistical significance on a continuum rather than as some binary choice. I concur. Many of these researchers also recommend reporting CIs in addition to significance tests, to which I also concur. I return to these CIs in Section 6.8.

5.8.5 Statistical Significance Tests for Population Data

Researchers often analyze population data nowadays, akin to our data on the 240 Norwegian soccer players. Adhering to the principles of this chapter — random sampling, repeated sampling, the normal distribution, CIs, and sample-to-population inference — doing a significance test on population data seems a bit like carrying coals to Newcastle: When analyzing the population, there is no random chance variation between the sample and the population to study! Using significance tests on population data, which appears to be the norm nevertheless, has to be justified on other grounds. The first is to think of the population under study as a random sample from some imaginary *superpopulation*. We may for example think of our 240 soccer players as a random sample of all the players appearing in the top-tier soccer league throughout history.[34] The second option is to make an inference towards some *process* rather than towards a superpopulation. That is, we test for significance to get support for the (underlying) process being responsible for bringing about the association between x and y. A significant result thus implies that the process we shed light on in our population data might be generalizable to other situations and populations.[35]

That said, the process of interest might be contingent on the population under study. Generalization might be dubious in such cases.

5.8.6 Statistical Significance Tests for Non-Random Samples (i.e., Convenience Samples)

Researchers analyze samples not meeting the random draw criteria in many real-life situations. We may call such data "convenience samples." They include persons participating in an RCT and all sorts of on-site samples: students on campus on a particular day, patients at a hospital in a given week, and persons visiting a park on a Sunday. Our data on Christmas beers and our data on the students in the anchoring experiment fall into this convenience category. The frame of reference for significance tests on convenience samples should arguably be a process or a superpopulation – and not a population. Yet prevailing practice among most present-day researchers, at least as far as I can tell, is to do significance tests on non-random samples as if they in fact were random samples from well-defined populations. Berk (2004) and White and Gorard (2017) are not happy about this practice, and I think they have a valid point.

5.8.7 Significance Tests: A Swan Song?

The NHST framework has several shortcomings. In essence, a significance test answers the wrong question. We want the probability of the alternative hypothesis being correct; the probability of x having an effect on y in the population. A significance test does not give us that. Nor does it tell us anything about the probability of the null hypothesis being correct; the probability of x having no effect on y in the population. What we get instead is the probability of finding an effect of x on y in our sample, or a larger one, if there is no such effect in the population. This is a poor equivalent at the best of times. Yet it is the best one there is unless one is a fan of Bayesian statistics. That is a story for another day, however.

These remarks aside, significance tests will not leave academia anytime soon. Furthermore, I am not among the scholars advocating such an abandoning.[36] Yet there is room for improvement, especially regarding the tendency to use p-values below 5 percent as the dominant and die-hard corroboration of some more or less interesting x-y association in some population. The mandatory reporting of CIs is one step in the right direction, to which I return in Section 6.8.

5.9 Chapter Summary, Key Learning Points, and Further Reading

This chapter has been about inferential statistics, that is, the process of making an inference about what happens outside of our data based on what we find inside our data. More may be said on this topic; see the further reading paragraph below. Yet if you got the gist of what took place in this chapter, you are well-prepared for making valid inferences in your own statistical work. Below follows some key learning points:

- A population is a number of units (e.g., people, counties, countries, firms, cars, wines, transactions, stocks etc.) having at least one common feature or trait.
- A sample is some proportion or fraction of a typically larger population.

- Random sampling is the way to make sure a sample becomes representative of a larger population. Random sampling ensures that a sample typically gets to be a miniature-model of the full-scale population. The CLT proves this.
- Provided a random and thus representative sample, we can make approximate descriptions of an unknown population based on exact descriptions of a sample from this population.
- 95 percent CIs express the approximate descriptions or inferences about a population.
- 95 percent CIs build on the principles of random sampling, on repeated sampling, and on the shape and the variation of the normal distribution.
- Roughly speaking, a significance test is a means to find out if a statistical association between x and y in a sample is "large enough," given a predetermined level of confidence (often 95%), to warrant attention in the population.
- Weak statistical sample-association → random chance might have caused it → a large p-value → we dare not reject the null hypothesis saying that x and y are unassociated in the population → we keep the null hypothesis suggesting that x and y are unassociated in the population.
- Strong statistical sample-association → random chance has probably not caused it → a small p-value → we reject the null hypothesis saying that x and y are unassociated in the population → we get indirect support for the alternative hypothesis that x and y are associated in the population.
- Small difference in y between group A and B in a sample → random chance might have caused it → a large p-value → we dare not reject the null hypothesis saying that the groups are equal with respect to y in the population → we keep the null hypothesis suggesting that the groups are equal with respect to y in the population.
- Large difference in y between group A and B in a sample → random chance has probably not caused it → a small p-value → we get indirect support for the alternative hypothesis saying that the groups are unequal with respect to y in the population.
- *Note*! In practice, the assessment of statistical significance needs more careful consideration than the above and on-purpose simplified decision rules suggest.

Agresti (2018) and Freedman et al. (2007) are obvious sources for everything that has to do with inferential statistics or significance tests in the traditional random-sample-to-population manner. Wheelan (2013) explains inferential statistics in a non-technical and intuitive way for those finding formulas and equations troubling (like myself). Weisberg (2014) is a fascinating account of the history of inferential statistics; the same is Salsburg (2001). Lew (2012) tells you what you need to know about p-values, and Kline (2020) sums up the NHST controversy. Berk (2004), Schneider (2013), and White and Gorard (2017) are good places to start for assessments of the NHST practice, whereas Frick (1998) summarizes the process-based significance testing approach. Reinhart (2015) is, among other things, an entertaining account of bad practice in inferential statistics in medicine.

5.10 Do-Files in Stata and Syntax-Files in SPSS

As before I assume you have read Sections 2.9, 3.9, and 4.9 before taking on the present section. The commands appear in plain text "outside" of do-files (Stata) or syntax-files (SPSS) to save space. As before I add some comments to the commands on occasion. I assume throughout that the "correct" data set is in memory to avoid unnecessary repetition.

5.10.1 Stata-Commands in Do-Files

Stata-output 5.1

```
sum goals if season == 0
```

The `if season == 0` command means that Stata selects a subset of the units in the data for analysis, as per the stated condition. In Stata, the `==` means "equal to" when used in `if`-statements.

Stata-output 5.2

```
sum goals
```

Stata-output 5.3

```
sum alch_perc
```

Stata-output 5.4

```
mean goals
```

Stata-output 5.5

```
sum hours_exer
mean hours_exer
```

Stata-output 5.6

```
prop sport_club
```

Stata-output 5.7

```
oneway hours_exer gender, t
```

Stata-output 5.8

```
ttest hours_exer, by(gender) une
```

The command-part `une` at the end is necessary when the SDs in the two groups, that is, male and female students in this case, are (very) unequal. Otherwise, it is redundant.

Stata-output 5.9

```
reg hours_exer i.gender
```

Stata-output 5.10

```
sum totgoal_mean
```

Figure 5.1

```
kdensity totgoal_mean, normal
```

Stata-output 5.11

```
sum goals
```

Note that the data set for Stata-output 5.11 is not on the book's website.

Stata-output 5.12

```
reg hours_exer i.gender
```

Stata-output 5.13

```
reg hours_exer i.status
```

Stata-output 5.14

```
tab sport_club gender, col chi2
```

Stata-output 5.15

```
tab sport_club gender, col chi2 exp
```

Stata-output 5.16

```
tab fitness_cen gender, col chi2
```

Stata-output 5.17

```
reg hours_exer i.gender
test i1.gender = 1.1
```

Replace `1.1` with a figure of your own choosing (e.g., 0.9) to test against other values for the null hypothesis.

Stata-output 5.18

```
reg hours_exer i.gender
```

Note that the data set for Stata-output 5.18 is not on the book's website.

5.10.2 SPSS-Commands in Syntax-Files

Stata-output 5.1

```
COMPUTE filter_$=(season = 0).
VARIABLE LABELS filter_$ 'season = 0 (FILTER)'.
VALUE LABELS filter_$ 0 'Not Selected' 1 'Selected'.
```

```
FORMATS filter_$ (f1.0).
FILTER BY filter_$.
EXECUTE.
DESCRIPTIVES VARIABLES=goals
  /STATISTICS=MEAN STDDEV MIN MAX.
FILTER OFF.
USE ALL.
EXECUTE.
```

The COMPUTE filter_$-command creates the "filter" selecting the subset of units in the data for analysis. The FILTER OFF-command turns the filter off again (not unsurprisingly!).

Stata–output 5.2

```
DESCRIPTIVES VARIABLES=goals
  /STATISTICS=MEAN STDDEV MIN MAX.
```

Stata–output 5.3

```
DESCRIPTIVES VARIABLES=alch_perc
  /STATISTICS=MEAN STDDEV MIN MAX.
```

SPSS–output 5.1 (Stata–output 5.4)

```
EXAMINE VARIABLES=goals
  /PLOT BOXPLOT STEMLEAF
  /COMPARE GROUPS
  /STATISTICS DESCRIPTIVES
  /CINTERVAL 95
  /MISSING LISTWISE
  /NOTOTAL.
```

Stata–output 5.5

```
DESCRIPTIVES VARIABLES=hours_exer
  /STATISTICS=MEAN STDDEV MIN MAX.
EXAMINE VARIABLES=hours_exer
  /PLOT BOXPLOT STEMLEAF
  /COMPARE GROUPS
  /STATISTICS DESCRIPTIVES
  /CINTERVAL 95
  /MISSING LISTWISE
  /NOTOTAL.
```

Stata–output 5.6

```
NPTESTS
  /ONESAMPLE TEST (sport_club) BINOMIAL(TESTVALUE=0.5
LIKELIHOODSUCCESSCATEGORICAL=LIST(1)
SUCCESSCONTINUOUS=CUTPOINT(MIDPOINT))
  /MISSING SCOPE=ANALYSIS USERMISSING=EXCLUDE
  /CRITERIA ALPHA=0.05 CILEVEL=95.
```

Stata-output 5.7

```
ONEWAY hours_exer BY gender
  /STATISTICS DESCRIPTIVES
  /MISSING ANALYSIS.
```

SPSS-output 5.2 (Stata-output 5.8)

```
T-TEST GROUPS=gender(0 1)
  /MISSING=ANALYSIS
  /VARIABLES=hours_exer
  /CRITERIA=CI(.95).
```

SPSS-output 5.3 (Stata-output 5.9)

```
REGRESSION
  /MISSING LISTWISE
  /STATISTICS COEFF OUTS CI(95) R ANOVA
  /CRITERIA=PIN(.05) POUT(.10)
  /NOORIGIN
  /DEPENDENT hours_exer
  /METHOD=ENTER gender.
```

Stata-output 5.10

```
DESCRIPTIVES VARIABLES=totgoal_mean
  /STATISTICS=MEAN STDDEV MIN MAX.
```

Figure 5.2 (Kernel density plot without normal curve)

```
GGRAPH
  /GRAPHDATASET NAME="graphdataset" VARIABLES=totgoal_mean MISS-
ING=LISTWISE REPORTMISSING=NO
  /GRAPHSPEC SOURCE=INLINE.
BEGIN GPL
SOURCE: s=userSource(id("graphdataset"))
DATA: totgoal_mean=col(source(s), name("totgoal_mean"))
GUIDE: axis(dim(1), label("totgoal_mean"))
GUIDE: axis(dim(2), label("Density"))
SCALE: linear(dim(2), min(-5))
ELEMENT: line(position(density.kernel.epanechnikov(totgoal_mean*1)))
END GPL.
```

Figure 5.2 (Histogram with normal curve)

```
FREQUENCIES VARIABLES=totgoal_mean
  /HISTOGRAM NORMAL
  /ORDER=ANALYSIS.
```

Stata–output 5.11

```
DESCRIPTIVES VARIABLES=goals
  /STATISTICS=MEAN STDDEV MIN MAX.
```

Note that the data set for Stata–output 5.11 is not on the book's website.

Stata–output 5.12

```
REGRESSION
  /MISSING LISTWISE
  /STATISTICS COEFF OUTS CI(95) R ANOVA
  /CRITERIA=PIN(.05) POUT(.10)
  /NOORIGIN
  /DEPENDENT hours_exer
  /METHOD=ENTER gender.
```

Stata–output 5.13

```
RECODE status (1=1) (ELSE=0) INTO bf_gf.
RECODE status (2=1) (ELSE=0) INTO cohabit.
REGRESSION
  /DESCRIPTIVES MEAN STDDEV CORR SIG N
  /MISSING LISTWISE
  /STATISTICS COEFF OUTS R ANOVA
  /CRITERIA=PIN(.05) POUT(.10)
  /NOORIGIN
  /DEPENDENT hours_exer
  /METHOD=ENTER bf_gf cohabit.
```

SPSS–output 5.4 (Stata–output 5.14)

```
CROSSTABS
  /TABLES=sport_club BY gender
  /FORMAT=AVALUE TABLES
  /STATISTICS=CHISQ
  /CELLS=COUNT COLUMN
  /COUNT ROUND CELL.
```

Stata–output 5.15

```
CROSSTABS
  /TABLES=sport_club BY gender
  /FORMAT=AVALUE TABLES
```

```
/STATISTICS=CHISQ
/CELLS=COUNT EXPECTED COLUMN
/COUNT ROUND CELL.
```

Stata-output 5.16

```
CROSSTABS
/TABLES=fitness_cen BY gender
/FORMAT=AVALUE TABLES
/STATISTICS=CHISQ
/CELLS=COUNT COLUMN
/COUNT ROUND CELL.
```

Stata-output 5.17

SPSS-output 5.3 reports the regression. Yet there is to the best of my knowledge no SPSS regression option available for testing against an alternative null hypothesis, such as a H_0-coefficient of 1.1.

Stata-output 5.18

```
REGRESSION
/MISSING LISTWISE
/STATISTICS COEFF OUTS CI(95) R ANOVA
/CRITERIA=PIN(.05) POUT(.10)
/NOORIGIN
/DEPENDENT hours_exer
/METHOD=ENTER gender.
```

Note that the data set for Stata-output 5.18 is not on the book's website.

5.11 Chapter Exercises with Solutions

Exercises:

Exercise 1

Look at the statistics outputs from Chapter 3 and onwards! On the occasions I mention a "clear" or a "marked" association between x and y or use some similar phrase, check that this refers to a p-value below 0.05 or 0.10. Furthermore, check that the p-value is above 5 or 10 percent whenever I claim that x and y are not associated or use some similar phrase indicating a no-association.

Exercise 2 (data: `student_exercise`, see appendix C of Chapter 2 for data documentation)

2a Use a multiple regression model and examine the following hypotheses:
   (1) Male students exercise more hours than female students.
   (2) Fitness center members exercise more hours than non-members.
   (3) Sports club members exercise more hours than non-members.
   (4) There is a U-association between student age and exercise hours.

2b Add the variable financial situation (econ) to the multiple regression in 2a and examine the following hypothesis:
    Students experiencing better financial situations exercise more hours than students experiencing poorer financial situations.

2c Examine the following hypothesis (by extending the model in 2b):
    The effect of fitness center membership on exercise hours is larger for male students than it is for female students.

Exercise 3 (data: stud_tourism, see appendix B of Chapter 4 for data documentation)

3a Controlling for book_time, destin, and type_trip, examine the following hypothesis:
    The effect of length of stay on total trip expenditures is larger than 22 Euros.

3b Examine the following hypothesis (by extending the model in 3a):
    There is a stronger association between length of stay and total trip expenditures for package trips than it is for non-package trips.

Exercise 4 (data: x-mas_beer, see Sections 2.2 and 2.3 for data documentation)

4a Use multiple regression and examine the following hypotheses:
    (1) Beers produced outside of Norway are less costly than Norwegian beers.
    (2) There is a positive association between a beer's alcohol level and price.

4b Examine the following hypothesis (by extending the model in 4a):
    The effect of alcohol level on price is dependent on production location.

Exercise 5 (data: soccer, see appendix B of Chapter 2 for data documentation)

5a Controlling for age, age-square, nation_dum, and match_tot, examine the following hypothesis:
    More match experience during the season has a positive association with yearly income.

5b Examine the following hypothesis (by extending the model in 5a):
    The effect of match experience during the season on income is dependent on whether a player has appeared on a national team or not.

Answers to exercises (in Stata only; see Section 5.10 for equivalent SPSS syntaxes):

Exercise 2 (data: student_exercise, see appendix C of Chapter 2 for data documentation)

2a Use a multiple regression model and examine the following hypotheses:
    (1) Male students exercise more hours than female students.
    (2) Fitness center members exercise more hours than non–members.
    (3) Sports club members exercise more hours than non–members.
    (4) There is a U–association between student age and exercise hours.
        The above alternative hypotheses imply the following null hypotheses, which we actually test:
    (1) Male students exercise for the same number of hours as female students.
    (2) Fitness center members exercise for the same number of hours as non-members.
    (3) Sports club members exercise for the same number of hours as non–members.
    (4) There is no U–association between student age and exercise hours.

The multiple regression is:

```
. reg hours_exer i.gender i.fitness_cen i.sport_club c.age##c.age

      Source |       SS           df       MS          Number of obs   =        644
-------------+----------------------------------       F(5, 638)       =      40.88
       Model |  1545.76779          5  309.153557       Prob > F        =     0.0000
    Residual |  4824.89176        638  7.56252627       R-squared       =     0.2426
-------------+----------------------------------       Adj R-squared   =     0.2367
       Total |  6370.65955        643  9.90771314       Root MSE        =       2.75


------------------------------------------------------------------------------
  hours_exer |      Coef.   Std. Err.      t    P>|t|     [95% Conf. Interval]
-------------+----------------------------------------------------------------
      gender |
        male |   1.113019   .2366447     4.70   0.000     .6483227    1.577716
             |
 fitness_cen |
         yes |   2.387821    .219089    10.90   0.000     1.957598    2.818043
             |
  sport_club |
         yes |   1.819841   .2728031     6.67   0.000      1.28414    2.355541
         age |  -.6959417   .2093595    -3.32   0.001    -1.107059   -.2848246
             |
 c.age#c.age |   .0117045   .0037339     3.13   0.002     .0043723    .0190368
             |
       _cons |   12.38328   2.808992     4.41   0.000     6.867295    17.89927
------------------------------------------------------------------------------
```

All *p*-values in the column P>|t| are below 0.01 or 1 percent (two-sided tests); we may thus reject all four null hypotheses. Note that the alternative hypotheses are directional, and that one-sided tests are thus to be preferred. I do not repeat the interpretations of the regression coefficients or R^2; cf. Chapters 3 and 4.

2b Add the variable financial situation (econ) to the multiple regression in 2a and examine the following hypothesis:

Students experiencing better financial situations exercise more hours than students experiencing poorer financial situations.

The null hypothesis, which we actually test, is:

Students experiencing better financial situations exercise for the same number of hours as students experiencing poorer financial situations.

The multiple regression is:

```
. reg hours_exer i.gender i.fitness_cen i.sport_club c.age##c.age econ

Output omitted
                                                               .

------------------------------------------------------------------------------
  hours_exer |      Coef.   Std. Err.      t    P>|t|     [95% Conf. Interval]
-------------+----------------------------------------------------------------
      gender |
        male |   1.113672   .2366494     4.71   0.000     .6489645    1.578379
             |
 fitness_cen |
         yes |   2.392139   .2191361    10.92   0.000     1.961823    2.822456
```

```
  sport_club |
         yes |    1.80324    .2733229     6.60   0.000     1.266517    2.33996
         age |   -.6716629   .2107957    -3.19   0.002    -1.085601   -.257724
             |
 c.age#c.age |    .0112805   .0037585     3.00   0.003     .0039001     .01866
             |
        econ |    .1598979   .1615867     0.99   0.323    -.1574092     .47720
       _cons |    11.8524    2.85981      4.14   0.000     6.236609    17.468
```

We have a directional alternative hypothesis, and the regression coefficient for financial situation variable is positive (econ = 0.16). We should divide the *p*-value by two
to get 0.162 (0.323/2 ≈ 0.162). Still, we cannot reject the null hypothesis at the 5 percent
level and we must thus keep it. There is no positive association between students' financial situation and hours of exercise in the student population.

2c   Examine the following hypothesis (by extending the model in 2b):
     The effect of fitness center membership on exercise hours is larger for male
     students than it is for female students.
The null hypothesis, which we actually test, is:
     The effect of fitness center membership on exercise hours is the same for male
     and female students.
The multiple regression is:

```
. reg hours_exer i.gender##i.fitness_cen i.sport_club c.age##c.age econ

      Source |       SS           df       MS       Number of obs   =       64
-------------+----------------------------------   F(7, 636)       =     30.8
       Model |  1613.52799         7   230.503999   Prob > F        =    0.000
    Residual |  4757.13156       636   7.4797666    R-squared       =    0.253
-------------+----------------------------------   Adj R-squared   =    0.245
       Total |  6370.65955       643   9.90771314   Root MSE        =    2.734


  hours_exer |      Coef.   Std. Err.      t    P>|t|     [95% Conf. Interval]
-------------+----------------------------------------------------------------
      gender |
        male |    .4437322   .3331824     1.33   0.183    -.2105385    1.09800
             |
 fitness_cen |
         yes |    1.936447   .2706074     7.16   0.000     1.405055    2.46783
             |
     gender#|
 fitness_cen |
    male#yes |    1.288566   .453623      2.84   0.005     .3977858    2.17934
             |
  sport_club |
         yes |    1.826239   .2719394     6.72   0.000     1.292231    2.36024
         age |   -.6520254   .2097496    -3.11   0.002    -1.063911   -.240139
             |
 c.age#c.age |    .0108858   .0037404     2.91   0.004     .0035409    .018230
             |
        econ |    .120016    .1613097     0.74   0.457    -.196748     .4367
       _cons |   11.88927    2.844102     4.18   0.000     6.304308    17.4742
```

The coefficient for the gender by fitness center interaction (gender#fitness_cen male#yes) is 1.29 and its *p*-value is 0.005. We therefore reject the null hypothesis. The effect of fitness center membership on exercise hours is larger for male students than it is for female students in the population, which by now should come as no surprise.

Exercise 3 (data: stud_tourism, see appendix B of Chapter 4 for data documentation)
3a   Controlling for book_time, destin, and type_trip, examine the following hypothesis:
      The effect of length of stay on total trip expenditures is larger than 22 Euros.
The null hypothesis, which we actually test, is:
      The effect of length of stay on total trip expenditures is not larger than 22 Euros.
The multiple regression and hypothesis test are:

```
. reg tot_spend los book_time i.destin i.type_trip

      Source |       SS           df       MS        Number of obs   =      444
-------------+----------------------------------     F(4, 439)       =    61.66
       Model |   56832535.8         4    14208134     Prob > F        =   0.0000
    Residual |   101160825        439   230434.681    R-squared       =   0.3597
-------------+----------------------------------     Adj R-squared   =   0.3539
       Total |   157993361        443   356644.156    Root MSE        =   480.04


   tot_spend |      Coef.   Std. Err.      t    P>|t|     [95% Conf. Interval]
-------------+----------------------------------------------------------------
         los |    27.60737    2.79562     9.88   0.000     22.11291    33.10184
   book_time |    5.941777   2.701041     2.20   0.028     .6331987    11.25035
             |
      destin |
 Beyond Nor..|     445.487   58.94681     7.56   0.000      329.6339      561.34
             |
   type_trip |
Package trip |   -74.39578   56.88282    -1.31   0.192    -186.1923     37.4007
       _cons |    20.01848   44.93898     0.45   0.656     -68.3038    108.3408
------------------------------------------------------------------------------

. test los = 22

 ( 1)   los = 22

       F( 1,    439) =      4.02
            Prob > F =     0.0455
```

With the *p*-value 0.045 that should be divided by two given the directional alternative hypothesis, we reject the null and gain support for the effect of length of stay on total trip expenditures being larger than 22 Euros. Does the same conclusion hold for a similar test against 23 Euros? Spoiler alert! It does not.
3b   Examine the following hypothesis (by extending the model in 3a):
      The association between length of stay and total trip expenditures is stronger for package trips than it is for non-package trips.
The null hypothesis, which we actually test, is:
      The association between length of stay and total trip expenditures is of the same magnitude for package trips and non-package trips.

The multiple regression is:

```
. reg tot_spend book_time i.destin i.type_trip##c.los

      Source |       SS           df       MS        Number of obs   =       444
-------------+----------------------------------   F(5, 438)       =     50.30
       Model |   57629190            5   11525838   Prob > F        =    0.0000
    Residual |  100364171          438  229141.943   R-squared       =    0.3648
-------------+----------------------------------   Adj R-squared   =    0.3575
       Total |  157993361          443  356644.156   Root MSE        =    478.69


------------------------------------------------------------------------------
   tot_spend |      Coef.   Std. Err.      t    P>|t|     [95% Conf. Interval]
-------------+----------------------------------------------------------------
   book_time |   5.767015   2.695084     2.14   0.033    .4701108    11.0639
             |
      destin |
  Beyond Nor..|  449.6119   58.82285     7.64   0.000    334.0017    565.22
             |
   type_trip |
Package trip | -227.8827   99.96782    -2.28   0.023   -424.3589   -31.4064
         los |  25.78721   2.953737     8.73   0.000    19.98195    31.5924
             |
  type_trip#|
       c.los |
Package trip |  15.83348   8.491685     1.86   0.063   -.8560302    32.52
             |
       _cons |  38.73494   45.92321     0.84   0.399   -51.52229    128.992
------------------------------------------------------------------------------
```

The package trip by length of stay interaction (`type_trip#c.los Package trip`) has a coefficient of 15.83 and its *p*-value is 0.063. Since the alternative hypothesis is directional, however, we should divide the reported *p*-value by two. In other words, we reject the null at the 5 percent level because the one-sided *p*-value is 0.032. The association between length of stay and total trip expenditures is stronger for package trips than it is for non-package trips in the population.

Exercise 4 (data: x-mas_beer, see Sections 2.2 and 2.3 for data documentation)

Note! The Christmas beer data are not a random sample from some well-defined population. Tests of significance should thus be justified with reference to a super-population or to a process responsible for generating the association between *x* and *y*. (I am not claiming that any of these justifications are feasible in the present context, but that is another matter.)

4a Use multiple regression and examine the following hypotheses:

(1) Beers produced outside of Norway are less costly than Norwegian beers.
(2) There is a positive association between beers' alcohol level and price.

The above alternative hypotheses imply the following null hypotheses, which we actually test:

(1) Beers produced outside of Norway cost the same as Norwegian beers.
(2) There is no association between beers' alcohol level and price.

The multiple regression is:

```
. reg price i.country alch_perc

      Source |       SS           df       MS        Number of obs   =        75
-------------+----------------------------------     F(2, 72)        =     23.57
       Model |  53.4957996          2  26.7478998     Prob > F        =    0.0000
    Residual |  81.7042004         72  1.13478056     R-squared       =    0.3957
-------------+----------------------------------     Adj R-squared   =    0.3789
       Total |       135.2         74  1.82702703     Root MSE        =    1.0653


       price |      Coef.   Std. Err.      t    P>|t|     [95% Conf. Interval]
-------------+----------------------------------------------------------------
     country |
Outside Norway | -.377654    .254615    -1.48   0.142    -.8852198    .1299117
   alch_perc |  .5360226   .0791915     6.77   0.000     .3781572     .693888
       _cons |  2.032752   .6621321     3.07   0.003      .712816    3.352688
```

The coefficient for the production location variable is negative and thus in line with the alternative hypothesis. The coefficient's *p*-value, however, is 0.142 – or 0.071 if we apply a one-sided test, as we should. We still cannot reject H_0; we get no support for our alternative hypothesis (at the 5 percent significance level). In contrast, the coefficient for alcohol level is positive, with a *p*-value less than 0.0001 (that we also should divide by two). We reject H_0 and gain indirect support for a positive association between alcohol level and price with reference to a superpopulation or a process responsible for generating the association.

4b Examine the following hypothesis (by extending the model in 4a):
   The effect of alcohol level on price is dependent on production location.
The null hypothesis, which we actually test, is:
   The effect of alcohol level on price is not dependent on production location.
The multiple regression is:

```
. reg price i.country##c.alch_perc

      Source |       SS           df       MS        Number of obs   =        75
-------------+----------------------------------     F(3, 71)        =     18.07
       Model |  58.5329942          3  19.5109981     Prob > F        =    0.0000
    Residual |  76.6670058         71  1.07981698     R-squared       =    0.4329
-------------+----------------------------------     Adj R-squared   =    0.4090
       Total |       135.2         74  1.82702703     Root MSE        =    1.0391


       price |      Coef.   Std. Err.      t    P>|t|     [95% Conf. Interval]
-------------+----------------------------------------------------------------
     country |
Outside Norway |  2.36294   1.292974    1.83   0.072    -.2151766    4.941056
   alch_perc |  .6978549   .1076187     6.48   0.000     .4832692    .9124405
             |
     country#|
 c.alch_perc |
Outside Norway | -.3338483  .1545717   -2.16   0.034    -.6420555   -.0256412
             |
       _cons |  .717435   .8877243     0.81   0.422    -1.052637    2.487507
```

The location by alcohol level interaction (country#c.alch_perc Outside Norway) has a coefficient of −0.33 and its *p*-value is 0.034 or 3.4 percent. We reject the null at the 5 percent level and get indirect support for the effect of alcohol level on price being dependent on production location with reference to a superpopulation or to a process responsible for generating the interaction effect.

Exercise 5 (data: soccer, see appendix B of Chapter 2 for data documentation)

Note! The soccer player data are a population. Tests of significance must therefore be justified with reference to a superpopulation or to a process responsible for generating the association between *x* and *y*. (I am not claiming that any of these justifications are feasible in the present context, but that is another matter.)

5a Controlling for age, age-square, nation_dum, and match_tot, examine the following hypothesis:

More match experience during the season has a positive association with yearly income.

The null hypothesis, which we actually test, is:

More match experience during the season has no association with yearly income.

The multiple regression is:

```
. reg inc_year c.age##c.age i.nation_dum match_tot match_ses
```

Source	SS	df	MS			
				Number of obs	=	24
				F(5, 234)	=	20.3
Model	3.4270e+11	5	6.8539e+10	Prob > F	=	0.000
Residual	7.9002e+11	234	3.3762e+09	R-squared	=	0.302
				Adj R-squared	=	0.287
Total	1.1327e+12	239	4.7394e+09	Root MSE	=	5810.

inc_year	Coef.	Std. Err.	t	P>|t|	[95% Conf. Interval]	
age	21800.79	8177.531	2.67	0.008	5689.797	37911.7
c.age#c.age	-446.0768	151.1103	-2.95	0.003	-743.7873	-148.366
nation_dum						
yes	61710.37	8836.135	6.98	0.000	44301.82	79118.9
match_tot	211.4482	74.90973	2.82	0.005	63.8645	359.031
match_ses	1058.521	482.7816	2.19	0.029	107.3672	2009.67
_cons	-229009.3	109487.6	-2.09	0.038	-444716.7	-13301.8

The coefficient for match experience during the season is positive (1,058), as per the alternative hypothesis. Its one-sided *p*-value, given the directional alternative hypothesis, is 0.029 divided by 2: 1.45 percent. We reject the null and get indirect support for more match experience during the season having a positive association with yearly income with reference to a superpopulation or to a process responsible for generating the association.

5b Examine the following hypothesis (by extending the model in 5a):

The effect of match experience during the season on income is dependent on whether a player has appeared on a national team or not.

The null hypothesis, which we actually test, is:

The effect of match experience during the season on income is not dependent on whether a player has appeared on a national team or not.

The multiple regression is:

```
. reg inc_year c.age##c.age match_tot i.nation_dum##c.match_ses
```

Source	SS	df	MS			
				Number of obs	=	240
				F(6, 233)	=	16.89
Model	3.4327e+11	6	5.7211e+10	Prob > F	=	0.0000
Residual	7.8945e+11	233	3.3882e+09	R-squared	=	0.3030
				Adj R-squared	=	0.2851
Total	1.1327e+12	239	4.7394e+09	Root MSE	=	58208

inc_year	Coef.	Std. Err.	t	P>|t|	[95% Conf. Interval]	
age	21329.65	8272.208	2.58	0.011	5031.766	37627.53
c.age#c.age	-438.8649	152.3969	-2.88	0.004	-739.1169	-138.613
match_tot	215.1831	75.59345	2.85	0.005	66.2491	364.1172
nation_dum						
yes	71550.92	25568.17	2.80	0.006	21176.57	121925.3
match_ses	1168.571	553.0545	2.11	0.036	78.94421	2258.198
nation_dum#						
c.match_ses						
yes	-452.1644	1102.18	-0.41	0.682	-2623.677	1719.348
_cons	-224170.1	110315.2	-2.03	0.043	-441512.7	-6827.381

The match experience by national team dummy (`nation_dum#c.match_ses yes`) has no significant effect, with a *p*-value of 0.682. We keep the null and get no indirect support for the effect of match experience during the season on income being dependent on whether a player has appeared on a national team or not.

Notes

1 There are exceptions in which the data are interesting in themselves. I return to these later in the section.
2 Many books on statistics more or less equate data with a sample and the world beyond the data with a population. Because modern-day statistical analysis to an ever-increasing extent pertains to data that are not samples in this traditional sense, I do not.
3 I return to this issue in Section 5.8.
4 The more general inference from the data to the world beyond the data is typically not probabilistic.
5 We assume each of the students contacted agreed to take part in the survey and (politely) answered every question in the questionnaire. This, of course, never happens in real life. Yet the failure to do so has no bearing on the main points and conclusions of this chapter if those who participate/answer and those who do not are roughly similar. If this is not the case, we have some more work to do, and I return to this in Section 6.4.
6 Not every time according to some natural law, but most times according to *a priori* known statistical "laws" or regularities. In practice, we often use more complex random-sampling

techniques to ensure representativeness (see, e.g., Bryman, 2016), but the basic idea mentioned here still applies.

7 A "large" sample typically refers to a sample with more than 120 units. For such a sample, the exact number is 1.96 SDs. For smaller samples, say 30 units, the analogous number is 2.042. I use 2.0 as an easy-to-remember approximation if not explicitly stated otherwise, but see Section 5.7 for more on this.

8 Statisticians also tell us that 68 percent of the beers lie in the interval mean ± one SD.

9 We get the standardized alcohol level in one beer by subtracting the mean alcohol level (8.187) from its actual alcohol level and dividing the result we get with the SD (1.566).

10 This imagined distribution carries a special name in statistics: the *sampling* distribution. The CLT proves that this distribution becomes close to normal for samples larger than 30 units or so.

11 The mean of the variable "Mean of total goals" if the repeated sampling continues for, say, 1,000 times approximates the unknown population mean. This thought experiment mean also carries a special name: the *sampling* mean.

12 There is in general much more variation among the units in a sample with respect to x (or y) than there is variation in the mean of x (or y) among samples in repeated sampling.

13 Thanks to Ulrik Berg Rian, Simen Kleven, and Marthe Sælebakke Stangnes for the permission to use the data they compiled from the Internet in January and February 2021.

14 We may have a hypothesis about how one variable, say y, is close to some population constant, but I do not consider this case.

15 I use "affect" rather than the more correct "is associated with" for presentational ease, but I emphasize that the latter is more correct in most situations.

16 I mentioned in passing in Section 1.1 that a statistical association in itself does not explain why we observe this association and that we need some theory, reason, or mechanism to help us here. Motivation and habit formation are these reasons in the two preceding examples.

17 I emphasize once more: If the data are a random sample from a population, we are generally not interested in the sample/data per se.

18 We omit the (theoretical) reason for the hypothesis in the final hypothesis specification because it typically appears in the text preceding this specification.

19 We try to falsify what we do not believe in rather than to verify what we believe in. We need not get into the details of this philosophical stance. The key point is that falsification is better proof than verification.

20 Note that rejecting H_0 is not the same as accepting H_1. Rejection of H_0 only supports H_1 indirectly.

21 The t-value for the mean gender difference in hours of exercise in the ANOVA case is -5.30; cf. Stata-output 5.8 or SPSS-output 5.2. The negative sign reflects that the male mean is being subtracted from the female mean.

22 The upcoming explications assume such a large sample if not explicitly stated otherwise.

23 I repeat: 1.96, rather than 2.00, is the strictly correct number for large samples.

24 In the ANOVA-case, as in Stata-output 5.7, the p-value appears under the column heading Prob > F. In SPSS, the p-value in the ANOVA-case appears under the Sig. column heading.

25 More generally: The p-value is the probability of finding the effect/association/difference in a sample, or a larger one, if there is no effect/association/difference in the population.

26 Noise is what I call random chance. In a related spirit, Reinhart (2015) calls the p-value a measure of surprise: The lower the p-value, the more surprising it is to get the sample result you get if the null hypothesis is correct in the population.

27 To be clear, this analysis tacitly presupposes an alternative hypothesis in which we have some reason to expect that the variables student status and exercise hours are associated.

28 The observant reader might have noticed that we actually had a directional hypothesis for the regression between exercise hours and gender. The correct p-value is thus < 0.0001 divided by two. In this case, however, such halving amounts next to nothing in practical terms.

29 The correct quantity should here be degrees of freedom and not sample size. For most practical purposes, however, this amounts to the same. More on degrees of freedom later in this section.

30 Imagine that all rows and columns should be a sum of 100 for a 2×2 table. If you then put the number 30 in the upper-left cell of the table, the numbers in the three remaining cells must clockwise become 70, 30, and 70. The only *free* cell is the first cell: 30. This table thus has *one* degree of freedom, and larger tables must consequently have more degrees of freedom.

31 The well-respected journal *The American Statistician* devoted an entire issue to this topic as late as 2019 (Vol 73, S1).

32 The *p*-value 0.171 is based on a so-called *F*-test (i.e., *F*-test = 1.88 in the Stata-output). I have not explicated the *F*-test, but it works rather similar to the *t*-test and the chi-square test in principle. For more on the *F*-test, see Sirkin (2005).

33 Similar reasoning applies to ANOVA and all bivariate techniques of associating *x* and *y*. I use the regression-case example simply because it is by now most familiar.

34 I agree with Berk (2004) who finds such reasoning far-fetched and tautological in many cases. One might perhaps also argue that treating a population as a sample in order to do significance tests resembles putting the cart in front of the horse.

35 The statistical significance of the age-square variable in Section 4.4 (*p*-value = 0.019) should thus be justified with respect to a process yielding a non-linear age-income association.

36 It deserves mentioning that White and Gorard (2017) end up on a negative note; they advocate stopping significance testing altogether. See also Gorard (2019).

6 Doing Quantitative Research
Some Tricks of the Trade

6.1 Introduction and Chapter Overview

Things seldom go as smoothly as they have done in this book in real-life statistical analysis. The reason is obvious: I made the required data preparation in advance. To shed more light on such preparation, Sections 6.2–6.4 handle various aspects of statistical analysis having mostly to do with data groundwork before taking on the later descriptive, associational, and inferential analyses. Since doing data preparation is the main purpose in these three sections, I include the statistical commands in the main text.

Sections 6.2 and 6.3 show how to create new variables based on existing variables, whereas Section 6.4 takes on the handling of missing data. Section 6.5 addresses outliers that typically are a part of both the initial data preparation and the final associational and inferential analyses.

Regression analysis is *the* working horse when analyzing variable associations in the behavioral and social sciences. If we are to trust the results of regression analysis, however, it has to meet certain conditions or assumptions. Section 6.6 is all about these assumptions of regression analysis.

In real-life statistical analysis, we often lack a yardstick for assessing whether a variable association is weak or strong – or when judging if a statistical difference between two groups is small or large. Such a lack-of-yardstick scenario is, in short, what Section 6.7 is about.

I have made several implicit choices regarding the presentation of results from associational statistical analyses in the book so far. Section 6.8 turns these implicit choices into explicit pieces of advice in terms of what you should do – or at least offer some deliberate thoughts – when presenting associational statistical results to your readers/audiences.

Section 6.9 summarizes the chapter and lists the key learning points and further reading. Section 6.10 contains the do-file commands and syntax-file commands from Section 6.6 and onwards. Section 6.11 contains the exercises and answers in the usual manner.

6.2 Creating, Recoding, and Labeling New Variables

6.2.1 Creating a New (Dummy) Variable

Research often involves strict categorical x-variables with many categories, say five or more. We then face a dilemma as researchers. On the one hand, we do not want to compare an excess number of categories or groups with respect to y; this is tedious for

DOI: 10.4324/9781003252559-6

researchers and audiences alike. We thus often collapse categorical variables into new variables having fewer categories.[1] On the other hand, we do not want to mix categories or groups not belonging together – as in mixing apples and oranges. Regrouping categories into fewer categories thus needs some kind of justification. Below I illustrate using the status variable in the student exercise data; see the frequency table in Stata output 6.1. (SPSS of course yields the same result, making it redundant here.)

```
. tab status

      status |      Freq.      Percent         Cum.
-------------+-----------------------------------
      single |        329        51.09        51.09
       bf/gf |        206        31.99        83.07
     cohabit |        109        16.93       100.00
-------------+-----------------------------------
       Total |        644       100.00
```

Stata output 6.1 Frequency distribution for the variable status in the student exercise data.

The student status variable has three categories: single (coded 0), boyfriend or girlfriend (coded 1), or cohabiting or married (coded 2). It is fair to say that the two latter categories have something in common that the first has not, namely a "significant other." It thus makes sense, at least in this pedagogical context, to create a new variable with two categories based on the original variable: single (coded 0) and non–single (coded 1). Stata output 6.2 presents the procedures to create, recode, and label this new dummy variable.

The first command-line in the do-file creates the new variable; status_dum. The second line recodes the new variable in the manner suggested above. The third and fourth lines label the new dummy,[2] whereas the fifth asks Stata for a frequency table. We note that the non–singles correctly amount to 315 students (206 + 109).

```
. gen status_dum = status

. recode status_dum 0 = 0 1/2 = 1
(status_dum: 109 changes made)

. label define status_duml 0 "single" 1 "non-single"

. label value status_dum status_duml

. tab status_dum

  status_dum |      Freq.      Percent         Cum.
-------------+-----------------------------------
      single |        329        51.09        51.09
  non-single |        315        48.91       100.00
-------------+-----------------------------------
       Total |        644       100.00
```

Stata output 6.2 An example of creating, recoding, and labeling into a new dummy variable.

SPSS-output 6.1 reports a similar procedure in SPSS. The first command-line in the syntax-file recodes the original variable into the new variable status_dum as per the instruction at the outset. The second to seventh line creates and attaches the values to the new variable, and line eight executes the syntaxes. The two last lines yield the frequency table for the new dummy.

```
RECODE status (0=0) (1 thru 2=1) (ELSE=SYSMIS) INTO status_dum.
VARIABLE LABELS
status_dum 'status_duml' .
VALUE LABELS
status_dum
0 'single'
1 'non-single' .
EXECUTE.
FREQUENCIES VARIABLES=status_dum
   /ORDER=ANALYSIS.
```

status_duml

		Frequency	Percent	Valid Percent	Cumulative Percent
Valid	single	329	51,1	51,1	51,1
	non-single	315	48,9	48,9	100,0
	Total	644	100,0	100,0	

SPSS-output 6.1 An example of creating, recoding, and labeling into a new dummy variable.

In these examples, we created and recoded a variable having three categories into a variable with two categories. The procedure is similar in the four to three (or two) category case and so on.[3] Many such category reduction processes are oftentimes necessary for a typical research project.

6.2.2 Creating a Logarithmic Variable

We examined the variable log of wine price in Section 4.4. Furthermore, the soccer player data contain the variable log of yearly income. The commands to create logarithmic variables are straightforward in Stata and SPSS. For example, the command to generate the log of wine price in a Stata do-file was:

```
gen log_price = ln(price)
```

Similarly, the command in a syntax-file in SPSS was:

```
COMPUTE log_price=LN(price).
EXECUTE.[4]
```

To create other logarithmic variables, we just use new names in the generate/compute expression and replace the variable name within the parenthesis with other names.

6.3 Creating a New Variable by Combining Existing Variables

The variable hours of exercising and the number of times of exercise per week appear in the student exercise data. We can easily create the variable exercise hours per workout by dividing the former by the latter. Stata output 6.3 presents the do-file command to create the new variable and associates it to the student gender variable by means of regression analysis in the usual manner. The first command-line creates the new variable of interest; ex_h_pw. The new variable has 51 missing values because 51 students reported zero hours of exercising. (We cannot divide zero by anything.) We note that male students on average have almost 0.16 hours longer workouts than female students. This difference is statistically significant below 0.00011, with a 95 percent CI from 0.076 to 0.237. SPSS-output 6.2 presents the syntax-command to produce the new variable (two lines) and the similar regression.

```
. gen ex_h_pw = hours_exer/times_exer
(51 missing values generated)

. reg ex_h_pw i.gender

      Source |       SS          df       MS          Number of obs   =       593
-------------+----------------------------------      F(1, 591)       =     14.50
       Model |  3.36324998         1   3.36324998     Prob > F        =    0.0002
    Residual |  137.035439       591   .231870455     R-squared       =    0.0240
-------------+----------------------------------      Adj R-squared   =    0.0223
       Total |  140.398689       592   .237159948     Root MSE        =   .48153


    ex_h_pw |      Coef.    Std. Err.       t     P>|t|     [95% Conf. Interval]
-------------+--------------------------------------------------------------------
     gender |
       male |    .1566541    .0411325     3.81    0.000     .0758705    .2374377
      _cons |    1.501151    .0247672    60.61    0.000     1.452509    1.549794
```

Stata output 6.3 Example of creating a new variable based on two existing ones and associating the new variable to student gender.

We just divided one variable by another. We may, of course, multiply, add, or subtract variables in the same way. For example, we may add the variables membership in a fitness center and membership in sports club to create a new variable having three categories: non-member of either, member of a fitness center *or* sports club, and member of a fitness center *and* sports club. Stata output 6.4 presents the do-file commands to create the new variable and its frequency distribution.

```
COMPUTE ex_h_pw=hours_exer / times_exer.
EXECUTE.


REGRESSION
  /MISSING LISTWISE
  /STATISTICS COEFF OUTS CI(95) R ANOVA
  /CRITERIA=PIN(.05) POUT(.10)
  /NOORIGIN
  /DEPENDENT ex_h_pw
  /METHOD=ENTER gender.
```

SPSS-output 6.2 Example of creating a new variable based on two existing ones and associating the new variable to student gender.

(Continued)

Model Summary

Model	R	R Square	Adjusted R Square	Std. Error of the Estimate
1	,155[a]	,024	,022	,48153

a. Predictors: (Constant), gender

ANOVA[a]

Model		Sum of Squares	df	Mean Square	F	Sig.
1	Regression	3,363	1	3,363	14,505	,000[b]
	Residual	137,035	591	,232		
	Total	140,399	592			

a. Dependent Variable: ex_h_pw
b. Predictors: (Constant), gender

Coefficients[a]

Model		Unstandardized Coefficients		Standardized Coefficients	t	Sig.	95,0% Confidence Interval for B	
		B	Std. Error	Beta			Lower Bound	Upper Bound
1	(Constant)	1,501	,025		60,610	,000	1,453	1,550
	gender	,157	,041	,155	3,809	,000	,076	,237

a. Dependent Variable: ex_h_pw

SPSS-output 6.2 (Continued)

```
. gen multi_memb = fitness_cen + sport_club

. label define multi_membl 0 "non-memb." 1 "memb. FC or SC" 2 "memb. FC and SC

. label value multi_memb multi_membl

. tab multi_memb

     multi_memb |      Freq.     Percent        Cum.
----------------+-----------------------------------
      non-memb. |        234       36.34       36.34
 memb. FC or SC |        354       54.97       91.30
memb. FC and SC |         56        8.70      100.00
----------------+-----------------------------------
          Total |        644      100.00
```

Stata output 6.4 Example of creating a new index-variable based on two existing variables.

The commands are now familiar. We note that the typical or mode student is a member of a fitness center or a sports club (55 percent), and that only a small minority (9 percent) is a member of both. The technical term for this new variable is an *index*.

SPSS-output 6.3 presents the analogs syntax-commands to produce the new index variable and its frequency distribution.

```
COMPUTE multi_memb=fitness_cen + sport_club.
VARIABLE LABELS
multi_memb 'multi_membl' .
VALUE LABELS
multi_memb
0 'non-memb.'
1 'memb. FC or SC'
2 'memb. FC and SC' .
EXECUTE.
FREQUENCIES VARIABLES=multi_memb
  /ORDER=ANALYSIS.
```

multi_membl

		Frequency	Percent	Valid Percent	Cumulative Percent
Valid	non-memb.	234	36,3	36,3	36,3
	memb. FC or SC	354	55,0	55,0	91,3
	memb. FC and SC	56	8,7	8,7	100,0
	Total	644	100,0	100,0	

SPSS-output 6.3 Example of creating a new index-variable based on two existing variables.

6.4 Missing Data and What to Do about Them

Missing data is the rule rather than the exception in quantitative research. Yet this is not a problem by necessity. Problems pile up when data are missing for *non*-random reasons we know nothing about, or when we do know about these reasons but lack the means to do something about them. Missing data typically come in two guises that also may interact: missing values or missing cases (i.e., units).[5] We take on these two issues in turn below.

6.4.1 Missing Values

We have analyzed data with complete information for all units so far in the book. The 75 beers in the Christmas beer data had complete information for all variables; the soccer player data had complete variable information for the 240 players. Likewise, all the students politely answered every question in our student data sets. We thus had *no missing* values. This almost certainly never happens in real life![6] I scrutinize the missing value phenomenon below for what probably is the most typical case in quantitative research, namely, that some non-trivial proportion of respondents did not answer one or several questions in a survey questionnaire.

The new student data, described in appendix A to this chapter, also concern lifestyle variables. The sample consists of 331 female students answering all survey questions short of one: the question on their weight. For this question, only 289 of the female students gave a valid answer. That is, only 87 percent answered the question on their weight. Stata-output 6.5 shows the descriptive statistics in this regard. (There is no need to show the exact same results in SPSS.)

Variable |	Obs	Mean	Std. Dev.	Min	Ma:
weight_kg |	289	66.56125	10.11237	42	11

Stata-output 6.5 Descriptive statistics for the variable weight_kg in the female student weight data.

The average female student in the sample weighs 66.6 kilograms (1 kg = 1.20 pounds/lbs.). The range is 68 kg (110 − 42 = 68). Is this mean value a precise estimate of the unknown weight for the average female student in the population?[7] Perhaps yes and perhaps no. The answer might be yes if the female students not reporting their weight weigh the same as the female students reporting it on average.[8] We have two options if this is the case: We could analyze the reduced sample (n = 289), or we could replace the missing values with the mean weight (66.56 kg) and analyze the full sample (n = 331). The two strategies amount to the same.[9] But what if the 42 students not reporting their weight weigh noticeably more, or noticeably less than the 289 students reporting it? Then we have a non-representative sample of the population at our hands. Such a scenario has three consequences: First, the mean replacement of missing values makes no sense. Second, the 66.56 kg mean estimate for the population is *biased*. Third, and worse still, the results of all associational analyses involving the weight variable and some other variables are biased too. I explicate using the regression case below.

Say we want to regress weight on height and age in the usual manner. We then have several options on how to handle, or not to handle, missing values:

1a   Do nothing 1: Listwise deletion, that is, the default in most statistics programs
1b   Do nothing 2: Casewise deletion
2a   Replacing the missing value with the overall mean, median, or mode, that is, manual imputation
2b   Replacing the missing value with the overall mean, median, or mode (manual imputation) for a certain subgroup in the data
3   Replacing the missing value with a prediction from a regression model, that is, data-analytic imputation

Stata and SPSS use listwise deletion (1a) as default. That is, the regression uses only the units for which there are no missing values among the variables included in the analysis. For casewise deletion (1b), in contrast, any associational analysis tries to maximize the number of units on an analysis-by-analysis basis. Both (1a) and (1b), however, assume values are missing at random (MAR). The listwise deletion and hence reduced-sample regression appears in Stata-output 6.6 and SPSS-output 6.4. We find the expected and positive association between height and weight; taller female students weigh more than shorter female students on average. We also note a positive association between age and weight; older female students weigh more than younger female students on average.

```
. reg weight_kg height_cm age

      Source |       SS           df       MS            Number of obs   =        289
-------------+----------------------------------         F(2, 286)       =      30.41
       Model |  5164.97268         2   2582.48634        Prob > F        =     0.0000
    Residual |  24285.9333       286   84.9158506        R-squared       =     0.1754
-------------+----------------------------------         Adj R-squared   =     0.1696
       Total |  29450.906        288   102.26009         Root MSE        =      9.215


   weight_kg |      Coef.   Std. Err.      t    P>|t|     [95% Conf. Interval]
-------------+----------------------------------------------------------------
   height_cm |   .6217234   .0887508     7.01   0.000     .4470357    .7964111
         age |   .3304169   .0993607     3.33   0.001     .134846    .5259878
       _cons |  -45.91221   15.1129     -3.04   0.003    -75.65883   -16.16559
```

Stata-output 6.6 Female students' weight by their height and age. Linear regression.

```
REGRESSION
 /DESCRIPTIVES MEAN STDDEV CORR SIG N
 /MISSING LISTWISE
 /STATISTICS COEFF OUTS R ANOVA
 /CRITERIA=PIN(.05) POUT(.10)
 /NOORIGIN
 /DEPENDENT weight_kg
 /METHOD=ENTER height_cm age.
```

Descriptive Statistics

	Mean	Std. Deviation	N
weight_kg	66,5612	10,11237	289
height_cm	168,5538	6,11888	289
age	23,24	5,465	289

Model Summary

Model	R	R Square	Adjusted R Square	Std. Error of the Estimate
1	,419[a]	,175	,170	9,21498

a. Predictors: (Constant), age, height_cm

ANOVA[a]

Model		Sum of Squares	df	Mean Square	F	Sig.
1	Regression	5164,973	2	2582,486	30,412	,000[b]
	Residual	24285,933	286	84,916		
	Total	29450,906	288			

a. Dependent Variable: weight_kg

b. Predictors: (Constant), age, height_cm

SPSS-output 6.4 Female students' weight by their height and age. Linear regression.

(Continued)

Coefficients[a]

Model		Unstandardized Coefficients		Standardized Coefficients	t	Sig.
		B	Std. Error	Beta		
1	(Constant)	-45,912	15,113		-3,038	,003
	height_cm	,622	,089	,376	7,005	,000
	age	,330	,099	,179	3,325	,001

a. Dependent Variable: weight_kg

SPSS-output 6.4 (Continued)

Approaches (2a) and (2b) replace the missing value with a sample value – such as the mean. We typically also add a dummy for the valid versus missing values when using this strategy. Stata-output 6.7 shows the do-file commands and the results. Note that the dot (.) means missing value in Stata; hence . = 1.

```
. gen weight_miss = weight_kg
(42 missing values generated)

. recode weight_miss 0/120 = 0 .= 1
(weight_miss: 331 changes made)

. label define weight_miss1 0 "valid" 1 "missing"

. label value weight_miss weight_miss1

. tab weight_miss

weight_miss |      Freq.      Percent        Cum.
------------+-----------------------------------
      valid |        289        87.31       87.31
    missing |         42        12.69      100.00
------------+-----------------------------------
      Total |        331       100.00

. gen weight_m_f = weight_kg
(42 missing values generated)

. recode weight_m_f . =  66.561246
(weight_m_f: 42 changes made)

. reg weight_m_f height_cm age i.weight_miss

      Source |       SS           df       MS       Number of obs   =        331
-------------+----------------------------------   F(3, 327)       =      20.36
       Model |  4634.71619         3   1544.9054   Prob > F        =     0.0000
    Residual |  24816.1898       327  75.8904886   R-squared       =     0.1574
-------------+----------------------------------   Adj R-squared   =     0.1496
       Total |  29450.9059       330  89.2451695   Root MSE        =     8.7115


  weight_m_f |      Coef.   Std. Err.      t    P>|t|     [95% Conf. Interval]
-------------+----------------------------------------------------------------
   height_cm |   .5549293   .0792041     7.01   0.000     .3991155    .7107432
         age |   .3033273   .0896674     3.38   0.001     .1269296     .479725
             |
 weight_miss |
     missing |   .4172449   1.441124     0.29   0.772     -2.4178     3.25229
       _cons |   -34.0242   13.49846    -2.52   0.012    -60.57898   -7.469416
------------------------------------------------------------------------------
```

Stata-output 6.7 Construction of missing value dummy, construction of new weight variable, and linear regression of female students' weight by their height, age, and a missing value dummy.

We note that the analysis uses the full sample and that the regression coefficients have changed very little in magnitude (0.62 versus 0.55 and 0.33 versus 0.30). Furthermore, the coefficient for the missing value dummy is not significant. We usually interpret both results as good signs and indirect support for that the missing data problem is not severe.

SPSS-output 6.5 presents the analogous syntaxes and output, but I have omitted some of the results in the output to make it more compact.

```
RECODE weight_kg (SYSMIS=1) (0 thru 120=0) (ELSE=Copy) INTO weight_miss.
VARIABLE LABELS
weight_miss 'weight_miss1' .
VALUE LABELS
weight_miss
0 'valid.'
1 'missing' .
FREQUENCIES VARIABLES=weight_miss
    /ORDER=ANALYSIS.
```

weight_miss1

		Frequency	Percent	Valid Percent	Cumulative Percent
Valid	valid.	289	87,3	87,3	87,3
	missing	42	12,7	12,7	100,0
	Total	331	100,0	100,0	

```
RECODE weight_kg (SYSMIS=66.561246) (ELSE=Copy) INTO weight_m_f.
REGRESSION
    /MISSING LISTWISE
    /STATISTICS COEFF OUTS R ANOVA
    /CRITERIA=PIN(.05) POUT(.10)
    /NOORIGIN
    /DEPENDENT weight_m_f
    /METHOD=ENTER height_cm age weight_miss.
```

Coefficients[a]

Model		Unstandardized Coefficients B	Std. Error	Standardized Coefficients Beta	t	Sig.
1	(Constant)	-34,024	13,498		-2,521	,012
	height_cm	,555	,079	,356	7,006	,000
	age	,303	,090	,172	3,383	,001
	weight_miss1	,417	1,441	,015	,290	,772

a. Dependent Variable: weight_m_f

SPSS-output 6.5 Construction of missing value dummy, construction of new weight variable, and linear regression of female students' weight by their height, age, and a missing value dummy.

We make an informed guess on what the missing value most likely is when replacing missing values with a mean, median, or mode. Option (3) uses regression analysis to make an even better guess in this regard. That is, we find the *expected* weight for the female students not reporting their weight by means of a multiple regression analysis based on the variables with complete information. Next, we replace (or impute) this expected weight for the missing values and do the associational analysis for the full sample.[10]

There were not many missing values in our case. In many real-life situations, however, the problem is severe – as in when 40 percent of the sample has not answered the *y*-question of interest. Options (1) and (2) appear dubious at the outset in such cases, especially when *y* refers to questions of a sensitive nature (e.g., weight, psychological traumas, sexual behavior, alcohol consumption etc.). Assuming that respondents answering sensitive questions are similar to respondents not answering such questions might sound dubious, but this should be evaluated on a case-by-case basis. The main approach in the social and behavioral sciences, as far as I can tell, is listwise deletion. The risk we take when analyzing incomplete data – that is, potentially analyzing non-representative data – seems to be more acceptable than the risk introduced by using manual or data-analytic imputation techniques.

6.4.2 Missing Values for a Dummy/Categorical y-Variable

Missing values for a dummy/categorical *y* is analogous to the problem for a continuous *y*. Yet the options for replacement values are fewer and the mean replacement is a no-go. Mode replacement is popular, that is, changing the missing values to the mode of *y* in the reduced sample. Furthermore, data-analytic imputation (3) is still possible. As before, the pros and cons should be assessed in the individual case.

6.4.3 Missing Values for x-Variables

Missing values for *x*-variables is typically a smaller problem. Suppose age is the *x*-variable of interest, but that a sizable proportion of your sample for some unknown reason has failed to report it. A remedy is to create a new age category variable (cf. Section 4.4) including a missing-age category for respondents with missing values. This way, you get the opportunity to compare respondents with missing age information to other age groups with respect to *y*. The same principle applies to variables that are categorical at the outset, that is, to recode missing values into a separate (answer) category.

6.4.4 Missing Cases (i.e., Units)

This section has mostly concerned missing values for a *y*-variable, that is, female students not answering a survey question on their weight. This section has not been about when cases are missing from the data to begin with. A classic example is when a person for some reason chooses *not* to take part in a survey. As mentioned earlier, we have no worries if those choosing to participate are similar to those not choosing to participate on average. When the two groups differ, however, problems arise: We get to analyze a non-representative sample of the population. The upcoming example illustrates this.

Think back on our survey questionnaire data on exercise habits, where Stata–output 6.8 displays the students' gender distribution. We note a roughly 65:35 gender distribution in the sample, with female dominance. (There is no need to present the exact same results in SPSS.) But suppose for now that the true gender distribution in the population was 50:50 and that we through random sampling contacted an equal proportion of male and female students.[11] For some unknown reason, however, male students chose more often than female students to abstain from taking part in the survey – yielding the skewed 65:35 distribution. In short, we have a sample that is unrepresentative of the population with respect to the gender variable's distribution.

```
. tab gender

     gender |      Freq.      Percent        Cum.
------------+-----------------------------------
     female |        416        64.60        64.60
       male |        228        35.40       100.00
------------+-----------------------------------
      Total |        644       100.00
```

Stata-output 6.8 Frequency table for the variable gender in the student exercise data.

Stata-output 6.9 shows the mean and CI for the variable exercise hours per week. The mean is 4.78, with a CI from roughly 4.5 to 5.0 (as we already knew from Chapter 5). The key question now is: How does our sample's non-representative gender distribution affect the unknown mean of the exercise hours variable in the population?

Chapter 5 showed that male students exercised more hours per week than female students. We now have (fictitious) information saying that male students are under-represented in our sample. We can thus deduce that the mean estimate of 4.78 hours is downward-biased: If we had a 50:50 gender distribution in our sample in sync with the population distribution, that is, a sample including more male students, the estimate for the mean of the exercise variable in the population would have been larger.

```
. mean hours_exer

Mean estimation                      Number of obs    =        644

------------------------------------------------------------------
            |       Mean    Std. Err.     [95% Conf. Interval]
------------+-----------------------------------------------------
 hours_exer |    4.78028    .1240349      4.536717     5.023842
------------------------------------------------------------------
```

Stata-output 6.9 Mean and 95 percent CI for hours _ exer in the student exercise data.

Statistics programs use *sampling weights* to account for a sample that is not representative of its population. Stata-output 6.10 shows the results in our (fabricated) case. We note that the mean estimate for exercise hours in the population increases to 4.99 hours if we base our calculation on a constructed 50:50 gender distribution for the students in the sample. The CI follows suit. The similar results in SPSS appear in SPSS-output 6.6.[12]

The key point in this subsection is this: We use sampling weights to get a sample *more* representative of the unknown population from which it is drawn.[13] This way, our results for the sample get closer to the corresponding (unknown) population results.

```
. mean hours_exer [pw=weight]
```

```
Mean estimation                       Number of obs   =        644
```

```
-----------------------------------------------------------------
             |      Mean    Std. Err.     [95% Conf. Interval]
-------------+---------------------------------------------------
 hours_exer |    4.991471    .1395307       4.71748    5.265461
-----------------------------------------------------------------
```

Stata-output 6.10 Mean and 95 percent CI for hours _ exer in the student exercise data, with the weighting variable weight to "construct" a 50:50 gender distribution in the sample.

We also apply sampling weights to the analysis of variable associations – much in the same manner as we just did for descriptive analysis. Finally, the problem of missing values (e.g., not answering a question) and missing cases (e.g., not participating in a survey) might appear in combination. But let us not go there – at least not yet!

```
WEIGHT BY weight.
EXAMINE VARIABLES=hours_exer
   /PLOT BOXPLOT STEMLEAF
   /COMPARE GROUPS
   /STATISTICS DESCRIPTIVES
   /CINTERVAL 95
   /MISSING LISTWISE
   /NOTOTAL.
```

Descriptives

			Statistic	Std. Error
hours_exer	Mean		4,9915	,12896
	95% Confidence Interval for Mean	Lower Bound	4,7382	
		Upper Bound	5,2447	
	5% Trimmed Mean		4,8078	
	Median		4,5000	
	Variance		10,711	
	Std. Deviation		3,27275	
	Minimum		,00	
	Maximum		16,00	
	Range		16,00	
	Interquartile Range		3,50	
	Skewness		,854	,096
	Kurtosis		,817	,192

SPSS-output 6.6 Mean and 95 percent CI for hours _ exer in the student exercise data, with the weighting variable weight to "construct" a 50:50 gender distribution in the sample.

6.5 Outliers: When Too Much Information Causes Trouble

The problem in Section 6.4 was lack of information, that is, the failure to answer a survey question or not to respond to a survey at all. The problem in this section is the reverse in a sense; we have too much information. If this sounds cryptic, I promise to make it more translucent during the course of the section. For starters, look at the data in Table 6.1. I have borrowed the data from Vries (2019); a book I also recommend as a user-friendly introduction to thinking critically about statistics.

The data have two variables: a Gini score and a homicide rate. The units are 23 rich countries. The Gini score measures the economic inequality in a country: Larger scores imply that more of the total incomes and fortunes belong to fewer people in a relative sense. A theoretical Gini score of 100 implies that one person earns/owns everything. In contrast, a score of zero suggests that everyone earns/owns equally much of everything. The homicide variable is the rate per 100,000 people, that is, a typical way of measuring some quantity at the country level.

Glancing through the data, we notice that many of the Gini scores lie in the 30 to 40 range. We also find many homicide rates around 1.0. Stata-output 6.11 presents descriptive statistics for these two continuous variables. The average Gini score is 32.23; the average homicide rate is 1.34. (The results are similar in SPSS of course.)

```
. sum gini homi

    Variable |        Obs        Mean    Std. Dev.        Min        Max
-------------+--------------------------------------------------------
        gini |         23    32.22913    4.375972      25.86      42.78
        homi |         23    1.343478    .8617246         .6        4.8
```

Stata-output 6.11 Descriptive statistics for gini and homiç

Table 6.1 Gini score of economic inequality and homicides per 100,000 people for 23 rich countries.

Country:	Gini score	Homicides per 100,000 people
Australia	34.94	1.0
Austria	30.25	0.7
Belgium	28.53	1.7
Canada	33.68	1.6
Czech Republic	26.63	1.0
Denmark	29.02	0.8
Finland	27.74	2.2
France	33.78	1.3
Germany	31.14	1.0
Greece	34.48	1.6
Ireland	32.30	1.1
Israel	42.78	2.0
Italy	34.41	0.9
Netherlands	28.73	0.9
Norway	25.86	0.6
Poland	33.25	1.1
Portugal	35.84	1.2
Slovak Republic	27.32	1.6
Spain	35.79	0.9
Sweden	26.81	1.0
Switzerland	32.72	0.7
UK	34.81	1.2
USA	40.46	4.8

Note. The data are called gini_hom.

The context for bringing up the data in Table 6.1 is the idea that economic inequality seems to be associated with many societal outcomes. In particular, many scholars within economics, sociology, and political science argue, yet not without controversy, that more economic inequality appears to be associated with many non-beneficial societal features – such as homicides. A regression analysis sheds light on this hypothesis. Stata-output 6.12 presents the results in this regard, whereas Figure 6.1 depicts it. We note a positive and significant regression coefficient for Gini (0.091; $p = 0.026$) yielding the upward-sloping regression line.[14] More economic inequality seems to bring about more homicides as per the theoretical expectation.

```
. reg homi gini

      Source |       SS           df       MS      Number of obs   =        2
-------------+----------------------------------   F(1, 21)        =      5.7
       Model |  3.49141264         1   3.49141264  Prob > F        =   0.026
    Residual |  12.8451106        21   .611671932  R-squared       =   0.213
-------------+----------------------------------   Adj R-squared   =   0.176
       Total |  16.3365232        22   .742569237  Root MSE        =   .7820

-------------------------------------------------------------------------------
        homi |      Coef.   Std. Err.      t    P>|t|     [95% Conf. Interval
-------------+-----------------------------------------------------------------
        gini |   .0910363   .0381042     2.39   0.026     .0117942    .170278
       _cons |  -1.590543   1.238847    -1.28   0.213    -4.166867    .985780
-------------------------------------------------------------------------------
```

Stata-output 6.12 Regression of homi by gini.

Now, look at the country all by itself in the upper right corner of the figure. This is the USA, and it stands out in two respects (cf. Table 6.1): First, it has a much higher homicide rate than any other country in the data. Second, it has the second-largest Gini score trailing only Israel. We call the USA an *outlier* in statistics lingo. By definition, an outlier is located far away from the bulk of the other data points. Outliers are not necessarily problematic by themselves in quantitative research. The problems start when outliers also are *influential*.

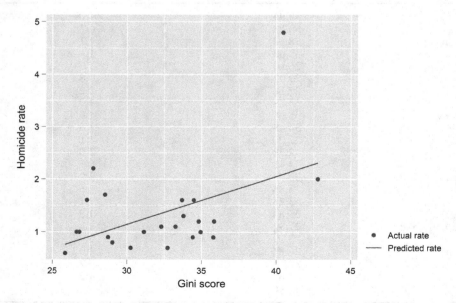

Figure 6.1 Graphical display of the regression in Stata-output 6.12.

Is the USA an influential outlier? We must do two statistical analyses to answer this question in its most fundamental sense. In the first, we use all the units — as in the regression in Stata-output 6.12 and Figure 6.1. In the second, we re-do the analysis without the potentially influential outlier: the USA. We then ask if the results of these two analyses differ. If not, we claim that the outlier is not influential. If yes, we conclude that the outlier is indeed influential. Stata-output 6.13 and Figure 6.2 show the results we need to make this comparative assessment.[15]

```
. reg homi gini if homi < 4

      Source |       SS           df       MS            Number of obs   =       22
-------------+------------------------------            F(1, 20)        =     1.16
       Model |  .211457614        1   .211457614         Prob > F        =   0.2936
    Residual |  3.63445169       20   .181722585         R-squared       =   0.0550
-------------+------------------------------            Adj R-squared   =   0.0077
       Total |  3.84590931       21   .183138538         Root MSE        =   .42629

------------------------------------------------------------------------------
        homi |      Coef.   Std. Err.      t    P>|t|     [95% Conf. Interval]
-------------+----------------------------------------------------------------
        gini |   .0245638   .0227713     1.08   0.294    -.0229364     .072064
       _cons |   .4038831   .7310527     0.55   0.587    -1.121066    1.928832
------------------------------------------------------------------------------
```

Stata-output 6.13 Regression of homi by gini, without the USA.

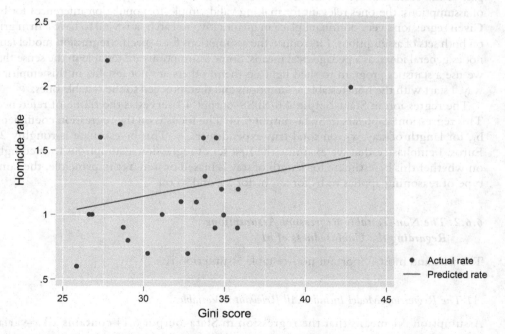

Figure 6.2 Graphical display of the regression in Stata-output 6.13.

The coefficient for Gini in Stata-output 6.13 is 0.025, that is, less than a third of the analogs coefficient in Stata-output 6.12. Consequently, the regression line is much less steep. Finally, the coefficient is not statistically significant ($p = 0.29$). The results of the two analyses differ, implying that the USA is an influential outlier. This country by itself, so to speak, "creates" a too-strong positive association between Gini scores and homicide rates. By removing the USA from the data, the positive association in the main disappears.[16]

The result above is intriguing in light of the research sparking the research question: Ignoring the USA, we find little evidence to support that economic inequality is a driver of homicide rates. We could even say that our first analysis was based on "too much information" as in too many countries. In most real-life analyses, we have to consider multiple outliers and not just one. The basic problem remains in the multiple case, however; what changes is that we often use data-analytic procedures to identify influential outliers. The visual approach, in other words, only works for small data sets. See the further reading section at the end of the chapter for more on how to handle multiple outliers.

6.6 The Assumptions of Regression Analysis

6.6.1 Preliminaries

When we for a sample associate x_1 and y by means of a regression while simultaneously controlling for x_2, x_3, and so on, we want to obtain the precise and unbiased estimate of x_1's effect on y in the population.[17] This effect is the regression coefficient or b_1. To be able to trust that b_1 is unbiased, the regression in question must meet some assumptions. The first part of this section addresses this set of assumptions. The second part addresses another set of assumptions: the ones relevant for making valid sample-to-population inferences for b_1. Given regression's very dominant place in quantitative research, it is vital to have a firm grip on both sets of assumptions. I introduce the assumptions for a specific regression model (and not as general ideas) as a pedagogical means. Some assumptions are testable in the sense that we use a statistics program to shed light on them; others are not testable in this empirical way. I start with the non-testable assumptions and then proceed to the testable ones.

The regression in Stata-output 4.6/SPSS-output 4.1 serves as the frame of reference. This regression reappears in Stata-output 6.14. The focus is on the regression coefficient, b_1, for length of stay, x_1, on total trip expenditures, y. This b_1-estimate is roughly 28 Euros. I emphasize that we examine the first set of regression assumptions to shed light on whether this b_1-estimate for length of stay is biased or not. Yet in principle, the same type of reasoning applies to b_2 for x_2, b_3 for x_3, and so on.

6.6.2 The Non-Testable Regression Assumptions
Regarding the Unbiasedness of b1

The first and most important non-testable assumption is:

A1 The Regression Model Includes All Relevant x-Variables

Assumption A1 means that the regression in Stata-output 6.14 contains all x-variables explaining variation in total trip expenditures. If this is the case, the entering of an additional x-variable will not cause length of stay's b_1 to change in magnitude. It is thus unbiased. The practical problem is that we can never be sure a regression actually contains all relevant x-variables. We can always think of some x_n we do not have in our data, and we cannot know for sure what would have happened to the b_1-estimate if this x_n was added to the regression: It could have increased, decreased, or remained unchanged. The b_1-estimate is thus potentially biased. (If we had x_n in our data, we would just have entered it to check!) The A1-assumption is not testable because it is impossible to control for x-variables that are not present in the data. At the end of the day, we can only hope our regression model contains the *most* relevant x-variables.[18] The regression in Stata-output 6.14 in all likelihood lacks relevant x-variables, such as students' income, students' savings, or their parents' income/savings. This omission might cause the b_1-estimate of 28 Euros to be biased.

```
. reg tot_spend los book_time i.destin i.type_trip

      Source |       SS           df       MS            Number of obs   =       444
-------------+----------------------------------         F(4, 439)       =     61.66
       Model | 56832535.8          4    14208134         Prob > F        =    0.0000
    Residual | 101160825         439  230434.681         R-squared       =    0.3597
-------------+----------------------------------         Adj R-squared   =    0.3539
       Total | 157993361         443  356644.156         Root MSE        =    480.04


-------------------------------------------------------------------------------------
   tot_spend |      Coef.   Std. Err.      t    P>|t|     [95% Conf. Interval]
-------------+-----------------------------------------------------------------------
         los |   27.60737    2.79562     9.88   0.000     22.11291    33.10184
   book_time |   5.941777   2.701041     2.20   0.028     .6331987    11.25035
             |
      destin |
Beyond Nordic |   445.487   58.94681     7.56   0.000     329.6339      561.34
             |
   type_trip |
Package trip | -74.39578   56.88282    -1.31   0.192    -186.1923     37.4007
       _cons |  20.01848   44.93898     0.45   0.656     -68.3038    108.3408
-------------------------------------------------------------------------------------
```

Stata-output 6.14 Total trip expenditures by independent variables for the student tourism data.

A2 The Regression Model Omits all Non-Relevant x-Variables

Textbooks always mention this assumption, although any violation of it typically is unproblematic. That said, something is often not right when a regression model contains more x-variables not contributing with significant effects than x-variables having significant effects. The A2-assumption is not testable for the same reason that the A1-assumption is non-testable. Furthermore, a non-significant effect of b_1 does not necessarily imply that x_1 is non-relevant.[19] In the present case, however, most of the x-variables contribute with significant effects that also make a certain kind of sense from a theoretical point of view. This is generally reassuring.

6.6.3 The Testable Regression Assumptions Regarding the Unbiasedness of b_1

There are four testable assumptions addressing the unbiasedness of b_1. These are:

B1 The linearity assumption
B2 The additivity assumption
B3 No influential outliers
B4 No (perfect) multicollinearity

I address them in turn below:

B1 The Linearity Assumption

Regression and linear regression often mean the same thing. Yet regression is the more general term because it says nothing about how x_1 and y are associated. Linear regression, in contrast, says that the association between x_1 and y is linear. As such, the linearity assumption applies to linear regression only. The typical way of checking if x_1 is linearly related to y is by means of a visual inspection. Figure 6.3 is a scatterplot smoother showing the four associations in Stata–output 6.14. The smoother puts no *a priori* restrictions, linear or otherwise, on the association between the four x-variables and total trip expenditures. Instead, the smoother lets the data speak for themselves. (Note that computation of smoothing plots is time–consuming for large data sets.)

The linearity assumption pertains only to length of stay and booking time; it does not apply to the two dummies. We see that length of stay and booking time have

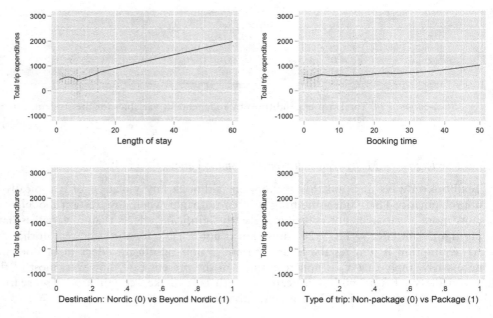

Figure 6.3 Multivariable scatterplot smoother visualizing the associations between the variables in Stata–output 6.14

approximately linear associations with total trip expenditures; the B1-assumption appears not to be violated. In cases where visual inspection reveals non-linear associations between the x-variables and y, that is, a violation of B1, we should employ the various strategies for handling non-linearity presented in Section 4.4.

B2 The Additivity Assumption

The B2-assumption says there are no significant interaction effects present. That is, non-additivity implies significant effects of interaction terms, whereas additivity means no significant interaction effects. We already know from Section 4.5 that our regression violates the B2-assumption because we found a significant effect of the interaction between length of stay and trip destination; cf. Stata-output 4.11. That is, the b_1-estimate for length of stay was not of the same magnitude for the two trip destinations of interest, but larger for trips to destinations beyond the Nordic countries than for trips to the Nordic countries. The take-home message is that a regression model containing one or more interaction terms is the procedure to solve a violation of the B2-assumption in the same way a non-linear regression is the procedure to solve a violation of the B1-assumption.

B3 No Influential Outliers

Section 6.5 showed how one unit in the data – that is, the influential outlier USA – caused the regression coefficient for the Gini score to be much larger than the analogous coefficient based on an analysis in which the USA was discarded from the data. This is a classic violation of the B3-assumption. A common solution to a B3-violation is to remove the influential outlier(s) from the data. This procedure is often trouble-free if the influential outliers are few compared to the rest of the units in the data. Yet such removing should be assessed against the possibility of excluding units actually belonging to the data. For random samples in particular, any outlier-deletion strategy might cause a representative sample to become non-representative in the extreme case. The general approach in the regression case is to do two regressions; one with and one without the potentially influential outliers. You do not have a B3-problem if the two analyses tell the same story regarding the magnitude of b_1. If they do not, you might consider dropping the influential outliers from the data.

There are more than 400 students in the data responsible for Stata-output 6.14. A visual inspection to identify potential influential outliers is thus cumbersome.[20] Thankfully, there are data-analytic methods available for identifying influential outliers in large data sets, but going into these is beyond the scope of this book; see the section on further reading at the end of the chapter.[21] Finally, it is very important you tell your readers how an outlier problem was handled if there indeed was one.

B4 No (Perfect) Multicollinearity[22]

The B4-assumption is not difficult to comprehend despite its complicated name. Multicollinearity has to do with associations among the x-variables in the regression model and has nothing to do with y. Say that trips lasting for many days always went to destinations beyond the Nordic countries and that trips lasting for a few days always went to Nordic countries. Then it would be hard for the regression to find out if the large total expenditures of a particular trip were the result of (1) being a long-lasting trip, or (2) being a beyond-Nordic destination trip.[23] This potential association between x-variables is the concern of multicollinearity; too strong an association is a violation of the B4-assumption. To examine B4, we calculate the Variance Inflation Scores (VIFs). Stata-output 6.15 presents the VIFs for the regression in Stata-output 6.14. (They are of course similar in SPSS.)

One problem with the mean VIF score as a measure of possible multicollinearity is

```
. vif

    Variable |      VIF      1/VIF
-------------+----------------------
         los |     1.08    0.929346
   book_time |     1.14    0.876530
    1.destin |     1.58    0.632136
 1.type_trip |     1.42    0.703804
-------------+----------------------
    Mean VIF |     1.30
```

Stata-output 6.15 VIFs for the regression in Stata-output 6.14.

that there is no general agreement regarding what constitutes a too-high score. Some claim multicollinearity is problematic only if the mean VIF score exceeds 10; others are concerned above 2.5.[24] Personally, I lean towards the lower threshold levels in this regard, say 3 or 4. In this case, however, multicollinearity is of no concern regardless of the preferred threshold value: None of the VIFs exceeds 1.58.

In the presence of multicollinearity between, say, x_1 and x_2, the estimates for b_1 and b_2 get *unreliable* and *unstable*. It is therefore questionable if these estimates capture the unbiased effects of x_1 and x_2. Multicollinearity between x_1 and x_2 causes no problems for the other x-variables and b-estimates in the regression model.

There are three main ways of dealing with multicollinearity: (1) Increase the number of units in the analysis if possible. (2) Delete the x-variable(s) responsible for the problem if this does not violate the A1-assumption, or combine the x-variables causing the problem into an index. (3) Do nothing if the regression estimates of main interest make theoretical sense despite having inflated SEs!

6.6.4 The Testable Regression Assumptions Regarding the Statistical Significance of b_1

I mentioned the error term or *e* briefly throughout Chapter 4. Now I bring *e* center stage. The *e* captures random variability's and all unmeasured x-variables' effects on *y* in the population. This *e* has to meet three assumptions if we are to trust the significance test of b_1: homoscedasticity in error terms (C1), normally distributed error terms (C2), and uncorrelated error terms (C3). In practice, however, we examine C1 to C3 for the *residuals*, which we may think of as the sample-equivalent of the error terms. I address the three assumptions in turn below.

C1 Homoscedastic Residuals

The C1-assumption, despite its complicated name, just means that the spread around the regression line should be of roughly similar size for the various levels of x_1. Figure 6.4 sheds preliminary light on the C1-assumption visually. Clearly, the spread around the regression line is *not* equal across the range of the length of stay variable. The residuals are not homoscedastic; they are heteroscedastic. The C1-assumption appears to be violated.

The plot in Figure 6.4 does not take all x-variables in Stata-output 6.14 into account. For this reason, the jury is still out on the final verdict for the C1-assumption. The residuals-versus-predicted-values plot in Figure 6.5 accounts for all x-variables in Stata-output 6.14. The horizontal and dashed line (at Residuals = zero) may be thought of as the regression line for the combination of all x-variables in the model. We verify our initial and tentative conclusion: The spread of the residuals is not equal around the regression line. That is, we have heteroscedasticity and thus a violation of the C1-assumption.

A second way to examine homoscedasticity is by using a formal statistical test. Homoscedasticity expresses the null hypothesis in this regard, whereas heteroscedasticity is the alternative hypothesis. Stata-output 6.16 presents the result of this homoscedasticity test.

We reject the null and get indirect support for heteroscedasticity, with a p-value below 0.00001. The statistical test corroborates our visual inspection: The C1-assumption is violated. Not meeting the C1-assumption causes the SE of b_1 to be biased.[25] We do not know if the SE is biased upwards or downwards, but the latter is more likely and a more serious problem. Remember from Section 5.7: We get the t-value by dividing b_1 on its SE. A downward-biased SE thus yields an upward-biased t-value. Long story short: If the SE is downward-biased, the *correct* t-value is smaller than what the statistics output suggests. We therefore risk claiming that b_1 is statistically significant when it actually is not.

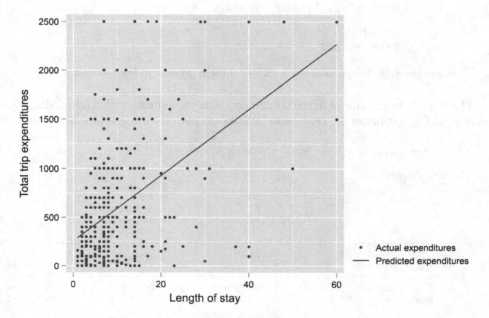

Figure 6.4 Scatterplot of correlation between length of stay and total trip expenditures, with regression line.

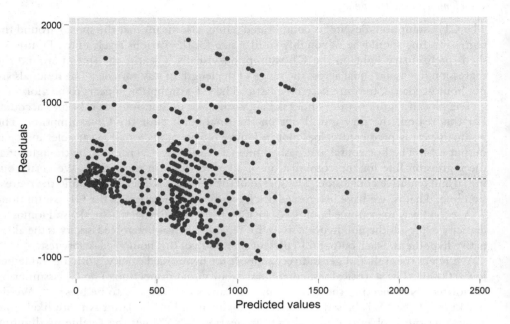

Figure 6.5 Residuals-versus-predicted-values plot for the regression in Stata-output 6.14.

```
. estat hettest

Breusch-Pagan / Cook-Weisberg test for heteroskedasticity
        Ho: Constant variance
        Variables: fitted values of tot_spend

        chi2(1)      =     88.60
        Prob > chi2  =    0.0000
```

Stata-output 6.16 Test of homoscedasticity for the regression in Stata-output 6.14.

The remedy in the face of heteroscedasticity is to compute so-called *robust* SEs. Stata-output 6.17 re-estimates the regression in Stata-output 6.14 using robust SEs.

```
. reg tot_spend los book_time i.destin i.type_trip, rob

Linear regression                          Number of obs   =        444
                                           F(4, 439)       =      55.00
                                           Prob > F        =     0.0000
                                           R-squared       =     0.3597
                                           Root MSE        =     480.04

-------------------------------------------------------------------------------
               |               Robust
    tot_spend  |     Coef.    Std. Err.      t    P>|t|    [95% Conf. Interval]
---------------+---------------------------------------------------------------
          los  |  27.60737   3.469799     7.96   0.000    20.78789    34.4268
    book_time  |  5.941777   3.410124     1.74   0.082    -.7604204   12.6439
```

Stata-output 6.17 Total trip expenditures by independent variables for the student tourism data, with heteroscedasticity-robust SEs.

(Continued)

destin						
Beyond Nordic	445.487	60.6321	7.35	0.000	326.3217	564.6522
type_trip						
Package trip	-74.39578	64.64582	-1.15	0.250	-201.4495	52.65797
_cons	20.01848	35.83057	0.56	0.577	-50.40229	90.43924

Stata-output 6.17 (Continued)

We note larger robust SEs for all x-variables compared with the plain vanilla SEs in Stata-output 6.14. Moreover, the enlarged SE of the booking time coefficient makes this coefficient insignificant at the 5 percent level ($p = 0.082$). The take-home lesson is that we should compute robust SEs whenever the C1-assumption is violated. It is worth mentioning that the presence of heteroscedastic error terms is a very typical scenario in most real-life applications of regression analysis in the social and behavioral sciences.[26]

C2 Normally Distributed Residuals

There is no need to show the normal distribution by now since we have seen it on several occasions. The C2-assumption unsurprisingly tells us that the distribution for the residuals should follow a normal distribution. Otherwise, the SEs will be biased – as in the case of a violation of the C1-assumption. Figure 6.6 shows the distribution for the residuals using a Kernel density plot alongside the normal distribution. The C2-assumption seems to be violated because the two curves do not align perfectly. We

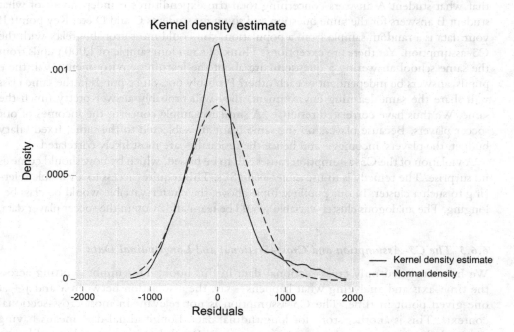

Figure 6.6 Kernel density plot and the normal distribution for the residuals based on the regression in Stata-output 6.14.

might also use a formal test to examine the C2-assumption. The results of such a test appear in Stata-output 6.18. The normal distribution is the null hypothesis, and the p-value below 0.000001 suggests a rejection of this null. We conclude once more that the residuals are not normally distributed.

```
. swilk r

              Shapiro-Wilk W test for normal data

   Variable |      Obs        W         V         z      Prob>z
------------+------------------------------------------------------
          r |      444    0.95017    15.063    6.485    0.00000
```

Stata-output 6.18 Test of normal distribution for the residuals for the regression in Stata-output 6.14.

A violation of the C2-assumption brings about biased SEs and hence significance tests that cannot be trusted, as mentioned earlier. Yet any such violation is problematic for small samples only – typically less than 120 units or so. The violation thus causes no harm for our analysis of 444 students.

C3 Uncorrelated Residuals

The C3-assumption often has to do with data collection. In random samples akin to what we have analyzed in this section, uncorrelated residuals follow by necessity from the random sampling procedure in most cases. In our tourism case, uncorrelated residuals mean that what student A answers concerning total trip expenditures is independent of what student B answers for the same question and so on for students C and D etc. Key point: If your data is a random sample from a population, you could most probably relax with the C3-assumption. Yet there are exceptions. Think of a random sample of 1,000 pupils from the same school answering a questionnaire about the learning environment. Will these pupils' answers be independent of each other? Probably not. Since pupils in the same class will share the same learning environment, they will probably answer pretty much the same. We thus have correlated residuals. A similar example concerns the incomes of our soccer players. Because players on the same team are subjected to the same, fixed salary budget, the players' incomes – and hence the residuals – are most likely correlated.

A violation of the C3-assumption causes SEs to be biased, which by now should come as no surprise. The remedy is to use *cluster-robust* SEs. This requires access to a variable referring to such a cluster. In our pupil example above, the cluster-variable would be class belonging. The analogous cluster-variable would be team affiliation in the soccer player data.

6.6.5 The C3–Assumption and Cross-Sectional and Longitudinal Data

We have analyzed only cross-sectional data in this book. This implies cutting across the time-axis and analyzing what happens – say, the association between x and y – at one given point in time. The C3-assumption is not relevant in most cross-sectional contexts. This is another story for longitudinal data. Longitudinal data means having repeated observations for the same unit over time. Panel data is a special kind of longitudinal data, as in when a random sample of respondents answers the same questionnaire once every year or so over a ten-year period. For such panel data, the answers given by respondent A at year 1 will most likely be correlated with the answers given by him/her at year 2, 3, and so on. The residuals are thus correlated. Since we do not analyze longitudinal data or panel data in this book, we need not get into statistical models for such data (but see the further reading section at the end of the chapter).

6.6.6 The Regression Assumptions: A Summary

The following regression assumptions ensure that the estimate of x_1's effect on y in the population, b_1, is unbiased:

A1 The regression model includes all relevant x-variables. If violation, consult Section 6.6
A2 The regression model omits all non-relevant x-variables. If violation, consult Section 6.6
B1 The linearity assumption. If violation, consult Section 4.4
B2 The additivity assumption. If violation, consult Section 4.5
B3 No influential outliers. If violation, consult Sections 6.5 and 6.6
B4 No (perfect) multicollinearity. If violation, consult Section 6.6

The remaining regression assumptions have to do with the significance test of b_1: We want to make sure such a significance test is to be trusted. The assumptions are:

C1 Homoscedastic residuals. If violation, compute robust SEs
C2 Normally distributed residuals. A violation is only problematic for small samples
C3 Uncorrelated residuals. If violation, compute cluster-robust SEs

6.6.7 The Linear Probability Model (LPM) and the Regression Assumptions

We used linear regression on a dummy y in Section 4.7; we called this the LPM. The LPM violates two regression assumptions by mathematical necessity: C1 and C2. In other words, such violations are typically not critical. A better reason for not preferring the LPM (as opposed to doing logistic regression) is that the LPM might yield predictions below zero and above one. Because a dummy y may take on only zero and one as values, such predictions are nonsensical. The size and severity of this problem should probably be assessed on a case-by-case basis.[27]

6.7 Effect Sizes

What is a strong association between two variables? What is a large mean difference between the two groups? Many people are for understandable reasons interested in the strength of an association or in the size of a mean group difference. I said earlier that we typically need some yardstick found in prior research to answer such questions. Often, however, we lack such a benchmark. This section takes on answering how-strong and how-large questions for bivariate variable associations when lacking relevant background knowledge for the assessment of the x-y association in question. I consider three of the most typical effect size scenarios: (1) a continuous y and continuous x, (2) a continuous y and dummy x, and (3) a dummy y and dummy x. I present the results in Stata only in this short section to avoid tedious repetition. The results are similar in SPSS.

6.7.1 A Continuous y and a Continuous x

The Pearson correlation coefficient or *r* was briefly mentioned in passing in Section 3.4. The *r* is the typical effect size measure in the case of a continuous *y* and a continuous *x*. Stata-output 6.19 shows the Pearson correlation coefficient for the association between total trip expenditures and length of stay. The analogous bivariate regression coefficient for length of stay is 33.50 (not shown).

```
. corr tot_spend los
(obs=444)

             | tot_sp~d       los
-------------+------------------
   tot_spend |  1.0000
         los |  0.4753    1.0000
```

Stata-output 6.19 The Pearson correlation coefficient for total trip expenditures and length of stay.

The correlation coefficient is just short of 0.48. This puts the association in the medium effect size category according to the rule-of-thumb display below:

Pearson correlation coefficient:	Effect size:
0.20 or −0.20	Small
0.50 or −0.50	Medium
0.80 or −0.80	Large

6.7.2 A Continuous y and a Dummy x

Stata-output 6.20 reports the mean of total expenditures for trips to the Nordic countries and trips beyond the Nordic countries. The results suggest that trips beyond the Nordic countries incur much larger expenditures than trips to the Nordic countries; almost 800 Euros as opposed to almost 270 Euros on average.

```
. oneway tot_spend destin, tab

            |       Summary of tot_spend
     destin |      Mean   Std. Dev.        Freq.
------------+------------------------------------
     Nordic |  269.32353   305.75106         170
  Beyond No |  797.88321   643.03929         274
------------+------------------------------------
      Total |  595.50676   597.19692         444

                   Analysis of Variance
    Source            SS         df      MS            F     Prob > F
------------------------------------------------------------------------
Between groups     29309241.5     1   29309241.5     100.67    0.0000
Within groups       128684119   442  291140.542
------------------------------------------------------------------------
    Total           157993361   443  356644.156

Bartlett's test for equal variances:  chi2(1) =   95.9710   Prob>chi2 = 0.00
```

Stata-output 6.20 Total trip expenditures by destination. One-way ANOVA.

Cohen's *d* is the typical effect size measure for a continuous *y* and a dummy *x*. Stata–output 6.21 reports Cohen's *d* and Hedges' *g*. Hedges' *g* is preferable to Cohen's *d* when two groups differ in size, which they clearly do in this case. Otherwise, their interpretations are similar.

```
. esize twosample tot_spend, by(destin) cohensd hedgesg

Effect size based on mean comparison

                                    Obs per group:
                                        Nordic =              170
                                Beyond Nordic =              274
-------------------------------------------------------------------
     Effect Size |    Estimate       [95% Conf. Interval]
-----------------+-------------------------------------------------
       Cohen's d |   -.9795863       -1.181037      -.7771348
      Hedges's g |    -.977923       -1.179032      -.7758153
-------------------------------------------------------------------
```

Stata-output 6.21 Test of effect size in Stata output 6.20.

The rule–of–thumb display below puts the mean difference in trip expenditures for the two groups in the large effect size category, that is, −0.98.

Cohen's *d*/Hedges' *g*:	Effect size:
0.20 or −0.20	Small
0.50 or −0.50	Medium
0.80 or −0.80	Large
1.20 or −1.20	Very large

6.7.3 A Dummy y and a Dummy x

Cohen's *w* is a "correlation coefficient" for a cross-tabulation, to simplify a bit. That is, Cohen's *w* expresses the effect size for an association between two categorical variables.[28] Stata–output 6.22 reports a cross-tabulation between choice of accommodation

```
. tab accom destin, col chi2

             |        destin
      accom  |   Nordic  Beyond No |      Total
-----------+----------------------+-----------
Commercial |       44         180 |        224
           |    25.88       65.69 |      50.45
-----------+----------------------+-----------
   Private |      126          94 |        220
           |    74.12       34.31 |      49.55
-----------+----------------------+-----------
     Total |      170         274 |        444
           |   100.00      100.00 |     100.00

        Pearson chi2(1) =  66.5150   Pr = 0.000
```

Stata-output 6.22 Accommodation type choice by trip destination. Cross-tabulation. N = 444.

type and trip destination. We note that choice of commercial accommodation is much more common for trips to destinations beyond the Nordic countries than it is for trips to the Nordic countries. The difference is 40 percentage points (65.69 − 25.88 ≈ 40). Stata-output 6.23 displays Cohen's w in this regard, which also is similar to the effect measure *phi* in the 2 × 2 cross-tabulation case.

```
. phi accom destin

            |          destin
     accom  |    Nordic  Beyond No |      Total
-----------+----------------------+----------
Commercial |        44        180 |        224
   Private |       126         94 |        220
-----------+----------------------+----------
     Total |       170        274 |        444

            Pearson chi2(1) =  66.5150    Pr = 0.000
phi = Cohen's w = fourfold point correlation = 0.3871    phi-squared = 0.149
```

Stata-output 6.23   Test of effect size in Stata output 6.22.

Cohen's w is almost 0.39, making it a medium-plus effect according to the rule-of-thumb display below:

Cohen's w:	Effect size:
0.10 or −0.10	Small
0.30 or −0.30	Medium
0.50 or −0.50	Large

6.8 How to Present and Communicate Statistical-Association Results

We typically do not copy-paste the output from Stata or SPSS directly into a thesis, a research paper, or a PowerPoint, as I mentioned in passing in Section 2.2. The next four subsections show some usual and formal ways of presenting statistical-association results in a thesis, research paper, or PP based on the analyses in Sections 6.7 (i.e., cross-tabulation and ANOVA) and 6.6 (i.e., multiple regression).

6.8.1 The Cross-Tabulation

Table 6.2 shows a typical way of presenting the results of a cross-tabulation. All numbers are extracted from Stata-outputs 6.22 and 6.23.

The main thing when presenting the results of Table 6.2 to readers and audiences is the difference in accommodation choice between the students traveling to Nordic or non-Nordic destinations. This 40-percentage points difference in choice of accommodation type (66 − 26 or 74 − 34) expresses the significant association between the variables. To sum up this association, we could write something like:

> The table shows that students on trips to non-Nordic destinations have a significantly larger probability ($p < 0.001$) of choosing commercial accommodation than

Table 6.2 Accommodation type choice by trip destination. Cross-tabulation. $N = 444$.

	Trip destination:		
Accommodation type:	Nordic trip	Beyond Nordic trip	Total
Commercial	26% (44)	66% (180)	50% (224)
Private	74% (126)	34% (94)	50% (220)
Total	100% (170)	100% (274)	100% (444)

Note. The numbers in parentheses are frequencies. Pearson chi-square $(1, n = 444) = 66.52$, $p < 0.0001$. Cohen's $w = 0.39$.

students on trips to a Nordic country. For trips to non-Nordic destinations, the percentage choosing commercial lodging is 66, whereas the analogous percentage is 26 for trips to the Nordic countries. This 40-percentage points difference might be characterized as medium-sized.

The association in Table 6.2 is statistically significant and suggests that the two groups of students – the Nordic country travelers and the beyond-Nordic country travelers – differ with respect to accommodation choice. In other cases, the x-y association of interest might not be statistically significant. Is there a point in reporting a cross-table for such a non-significant x-y association? The correct answer is, not for the first time, that it depends. The upcoming example illustrates the general idea in question.

Statements (A) and (B) below convey the *exact* same amount of information:

(A) 20% of the total sample preferred option i for y.
(B) 20% of the females in the sample, and 20% of the males in the sample, preferred option i for y.

Think of (A) as a frequency table and (B) as a cross-tabulation. A non-significant gender difference implies in a strict sense that a cross-tabulation contains no more information than a frequency table, as in (A) and (B). Given this strict interpretation, the cross-table is redundant for a non-significant x-y association; it contains the same information as the frequency table. That said, and continuing with the example, gender equality with respect to y might be an interesting research finding in itself in certain circumstances and thus worthy of being mentioned. Sometimes equality among groups is as interesting as non-equality – especially when this is unexpected.

6.8.2 The One-Way ANOVA

Table 6.3 shows a typical way of presenting the results of a one-way ANOVA. All numbers are extracted from Stata-outputs 6.20 and 6.21.

Table 6.3 Mean of total trip expenditures, in total and by trip destination. One-way ANOVA.

Variable: total trip expenditures	$N =$	*Mean*	*SD*
Trip destination: Nordic trip	170	269.32	305.75
Trip destination: Beyond Nordic trip	274	797.88	643.04
Total	444	595.51	597.20

Note. $F(1, 442) = 100.67$, $p < 0.00001$. Cohen's $d = -0.98$ and Hedges' $g = -0.98$.

The main thing to communicate to readers and audiences in Table 6.3 is the significant and very large mean difference in expenditures between the students traveling to the two destinations.[29] As a brief summary, we could write something along the following lines:

> The table shows that students on trips beyond the Nordic countries spend significantly more money ($p < 0.00001$) than students on trips to the Nordic countries. On average, the students on trips to non–Nordic countries spend 798 Euros, whereas students on trips to Nordic countries spend 269 Euros. This mean difference should be classified as large.

We might find no significant mean differences for other x-y ANOVA-associations. That is, the overall mean of y might be indistinguishable from the mean of y among the subgroups of interest. Does this scenario make the ANOVA-table redundant in the sense that it contains no more information than the mean of y? The answer has lots in common with the answer for the cross-tabulation in Table 6.2: Yes, the ANOVA-table is redundant in a strict sense. That said, mean equality in y among subgroups in the data might be an interesting research finding in itself, especially when this is unexpected.

6.8.3 The Multiple Regression

A typical regression table based on Stata-output 6.14 or SPSS-output 4.1 might look something like Table 6.4.

Three general comments are of note at the outset. First, regression outputs from statistics programs contain much more information than what goes into regression tables. Second, we always report SEs. Third, we typically add one, two, or three asterisks (*, **, or ***) to indicate each regression coefficient's statistical significance level.[30]

The main findings to communicate to readers and audiences are the effects of length of stay and trip destination. That said, all regression coefficients should most often be commented. We might write something like the following as a summary:

> Length of stay has the expected positive effect on total trip expenditures: A trip lasting for, say, seven days incurs on average 28 Euros more than a trip lasting six

Table 6.4 Total trip expenditures in Euros by independent variables. Multiple linear regression.

Variables:	B
Length of stay (in days)	27.61 (2.80)***
Booking time (in weeks)	5.94 (2.70)*
Trip destination (1 = Beyond Nordic country)	445.49 (58.95)***
Type of trip (1 = Package trip)	−74.40 (56.88)
Constant	20.02
R^2	0.36
$N =$	444

Note. Standard errors are in parentheses.
* $p < 0.05$; ** $p < 0.01$; *** $p < 0.001$ (two-sided tests)

days *ceteris paribus* ($p < 0.001$). A ten-day difference in length of stay thus amounts to almost 280 Euros, making length of stay an important determinant of trip expenditures. Trips booked, say, ten weeks in advance incur six Euros more on average than trips booked nine weeks in advance *ceteris paribus* ($p < 0.05$). Trips to non–Nordic countries entail on average 445 Euros more than trips to Nordic countries *ceteris paribus* ($p < 0.001$), making also trip destination an important determinant of trip expenditures. In contrast, package trips do not seem to incur fewer expenditures than non-package trips *ceteris paribus* ($p > 0.05$). R^2 suggests that the four independent variables in the regression model explain 36 percent of the variation in total trip expenditures.

The regression model in Table 6.4 has four *x*-variables. Models that are more complex involve more *x*-variables, interaction terms, and/or square variables. With respect to presentation, however, one more such variable just means one more row in the table.

6.8.4 *Presenting CIs*

A response to the NHST-controversy mentioned in Section 5.8 is the recommendation to report 95 percent CIs in all research settings. Table 6.5 reports such CIs based on Stata–output 6.14.

Readers and audiences should be told that the CI for length of stay's coefficient goes from 22 to 33 Euros. It is hence a rather precise estimate – and always greater than zero. The CI for the booking time coefficient does not include zero either, although it is close. The coefficient for destination has a very broad CI, but it is always greater than zero. Finally, the coefficient for the type of trip has a CI that includes zero, making it statistically insignificant at $p > 0.05$. Figure 6.7 is a visual display of the above-mentioned regression coefficients along with their CIs.

Table 6.5 Total trip expenditures in Euros by independent variables. Multiple linear regression, with 95 percent confidence intervals.

Variables:		95 percent confidence interval	
		Lower	Higher
Length of stay (in days)	27.61*** (2.80)	22.11	33.10
Booking time (in weeks)	5.94* (2.70)	0.63	11.25
Trip estimation (1 = Beyond Nordic country)	445.49*** (58.95)	329.63	561.34
Type of trip (1 = Package trip)	−74.40 (56.88)	−186.19	37.40
Constant	20.02		
R^2	0.36		
$N =$	444		

Note. Standard errors are in parentheses.
* $p < 0.05$; ** $p < 0.01$; *** $p < 0.001$ (two-sided tests).

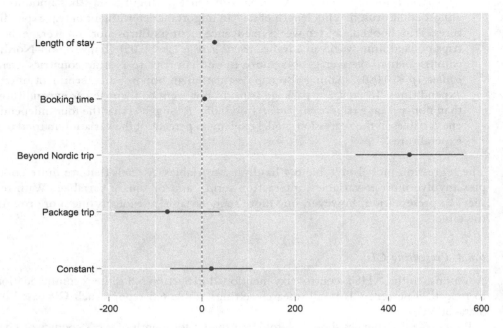

Figure 6.7 Regression coefficients and their CIs based on the regression in Table 6.5.

It is also possible to combine the graphical way of presenting regression results with the use of CIs. Figure 6.8 illustrates for the association between length of stay and total trip expenditures. We note the larger CIs – that is, the more variation – in predicted total expenditures for trips of longer duration. It is straightforward to expand Figure 6.8 with separate regression lines for different subgroups in the data, as we saw in Section 4.5 on interaction effects.

Finally, we might present graphs based on one-way ANOVAs that include CIs. Figure 6.9 illustrates the association between total trip expenditures and trip destination based on the numbers in Stata-output 6.20. We again note the huge difference in mean spending between the two groups, as well as two CIs that are far from overlapping.[31]

6.8.5 Presenting and Communicating Statistical-Association Results in Non-Academic Settings

Scholarly reports, theses, and research papers are one type of outlet for statistical results. The other main type is the non-academic setting – to coin a broad term. Below follows some practical pieces of advice on how to communicate statistical-association results in non-academic settings. That said, many of these prescriptions might come in handy in academic settings as well.

Tip 1: Know Your Audience – and Prepare Accordingly!

When preparing a presentation about the results of statistical associations, perhaps with the aid of PowerPoints,[32] it is often useful to have the typical spectator in your mindset.

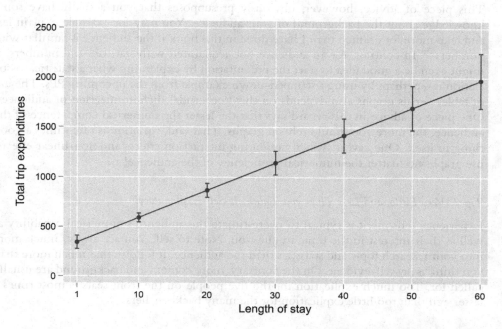

Figure 6.8 Effect of length of stay on total trip expenditures. Predictions with 95 percent CIs based on the regression in Table 6.5.

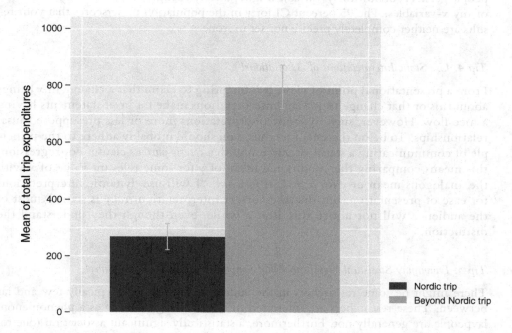

Figure 6.9 Means and 95 percent CIs for total trip expenditures by trip destination.

This piece of advice, however, obviously presupposes that you actually have some knowledge about the background of your audience. You may for example explain less and take more for granted than I have done in this book if the audience is familiar with numbers.[33] In contrast, for an audience not acquainted with statistics and numbers, it might even be a good idea to start the presentation by explaining what a statistical association is – perhaps by using a stripped-down example from the upcoming PPs. The use of tables versus graphs also depends on this (too crude) dichotomization of audiences. One piece of advice in this regard says that the lesser the numerical competence of the audience, the more you should rely on graphs. That said, variation is oftentimes a good thing in itself. One caveat: When presenting interaction effects and non-linear effects, use graphs no matter the numerical proficiency of the audience!

Tip 2: Very Little, and Less Than you Think, Is Self-Explanatory!

Experience tells me it is typical to overestimate the audience's numerical capability as well as their interest in the topic in question. Note to self: You are always much more into your research topic and statistics than the audience. It follows that much more than you think is *not* self-evident. On the contrary, more context and background are usually called for. Too much explication for the five people on the front seats is most times a lesser evil than too little explication for the many backbenchers.

Tip 3: Spend Time on the y-Variable!

It is always smart to allow the audience to get a firm grip on the characteristics of your y-variable.[34] Means, medians, mode (for categorical variables) are useful in order to let people get a feel of how the y-variable might possibly change with or without the "help" of any x-variables. The 95 percent CI for y in the population underscores that your results are neither completely precise nor set in stone.

Tip 4: Use Static Interpretations of Associations![35]

From a presentational point of view, it is tempting to claim that a change in x brings about this or that change in y. Such interpretations make for brief statements having a nice flow. However, such dynamic interpretations more or less presuppose causal relationships. To be on the safe side, thus, you should probably adhere to the principle of communicating a static association with a *ceteris paribus* clause. For regression, this means comparing the conditional mean of y for some relevant value of x with the analogous mean of y for $x + 1$. If you say, "I will use dynamic interpretations for ease of presentation, but that the correct interpretation *really* is static," most of the audience will not notice this after a while, even though they understand the distinction.

Tip 5: Downplay Statistical Significance and Emphasize Practical Significance!

There might be some researchers in the audience, but they are typically few and far between. These researchers are interested in statistical significance as a phenomenon; laypeople are generally not. Furthermore, a statistically significant association (due to, say, a large sample size) might be flat-out uninteresting from the practical point of view of most people in the audience. Hence, you should focus on practical significance; to what extent does a change in x appear to bring about a substantial change in y should be the main question. Remember that a practically significant association usually also is a statistically significant association.

6.9 Chapter Summary, Key Learning Points, and Further Reading

This chapter began with a coverage of some topics that come up most times in quantitative research projects before we take on the descriptive, associational, and inferential questions. In short, Sections 6.2 through 6.4 were about data preparation. In contrast, Section 6.5 was about outliers; a topic that is a part of the final associational and inferential analyses. Section 6.6 scrutinized the assumptions of regression analysis, whereas Section 6.7 was about effect sizes. Finally, Section 6.8 dealt with how to communicate statistical-association results to readers and audiences. Below follows some key learning points:

- Most quantitative research projects involve an initial phase of data preparation: recoding of variables, labeling of variables, and construction of new variables. This amounts to a lot of tedious work in most applications. A systematic streak is thus called for!
- Missing data are pervasive for data sets in the social and behavioral sciences, and they typically show up in two guises: missing values and missing cases (i.e., units). Missing values refer to when information is lacking on a variable for a particular unit – such as when a respondent does not answer a specific question in a survey questionnaire. Missing cases refer to when a unit that should be part of the data is not – as in when someone receiving a survey questionnaire chooses not to take part in the study.
- If data are missing (completely) at random, we might often do statistical analysis as if nothing has happened. However, if data are not missing at random, which arguably is the more typical case, we should address the missing data problem (if possible).
- Outliers are data points that in a physical sense are located far away from the bulk of other data points; cf. Figure 6.1. Outliers are influential if we by removing them from the data obtain a statistical result different from the one we get when keeping the outliers in the data. A common procedure is to discard influential outliers from the data, but such a strategy might have pitfalls.
- Regression analysis must meet a set of assumptions to make sure its results are correct in the sense that the sample regression coefficients are close to their unknown population counterparts. Some of these assumptions are testable in a statistics program; others are not. Section 6.6 guides you through the particulars in this respect.
- It is difficult to assess if a correlation is strong or weak – or if a mean group difference is large or small – when we have no prior information as to what constitutes "strong" or "large." Effect size measures help save the day in such circumstances. The correlation coefficient or r, Cohen's d, and Cohen's w are three typical measures of effect sizes.
- The output from statistics programs typically requires some tinkering before appearing as final tables or graphs in a thesis, a research paper, or a PowerPoint presentation. The characteristics of the audience govern the choice of how to present statistical-association results most effectively. As a broad generalization, tables tend to work well for numerically competent people, whereas graphs are more intuitive for the less numerically competent.

Further reading on missing data are Allison (2002) and Gorard (2020). Identifying out-liers is a vast topic, and Aggerwal (2017) provides a formal treatment, whereas Meule-man et al. (2015) is a concise and practically oriented explication in the regression context. Berry (1993), Meuleman et al. (2015), Allison (1999), and Berk (2004) illu-minate the assumptions of regression. Wooldridge (2010) covers all you need to know about analyzing longitudinal data and panel data. For more on effect sizes, see Cohen (1988) and Khamis (2008).

6.10 Statistical Commands: Do-Files in Stata and Syntax-Files in SPSS

As before I assume you have read Sections 2.9, 3.9, 4.9, and 5.10 before taking on this section. The commands appear in plain text "outside" of do-files (Stata) or syntax-files (SPSS) to save space. As before I add some comments to the commands on occasion. I assume throughout that the "correct" data set is in memory to avoid unnecessary rep-etition. Since the Stata and SPSS commands appear in the main text in Sections 6.1 to 6.5, the commands below refer to Section 6.6 and onwards.

6.10.1 Stata-Commands in Do-Files

Stata-output 6.14

```
reg tot_spend los book_time i.destin i.type_trip
```

Figure 6.3

```
mrunning tot_spend los book_time destin type_trip
```

Before doing the mrunning-command (i.e., the scatterplot smoother), you must first download it. In the Command-window, type findit mrunning and follow the in-structions after first clicking on SJ-5-3 gr0017.

Stata-output 6.15

```
reg tot_spend los book_time i.destin i.type_trip
vif
```

Figure 6.4

```
twoway (scatter tot_spend los) (lfit tot_spend los)
```

Figure 6.5

```
reg tot_spend los book_time i.destin i.type_trip
rvfplot, yline(0)
```

Stata-output 6.16

```
reg tot_spend los book_time i.destin i.type_trip
estat hettest
```

Stata-output 6.17

```
reg tot_spend los book_time i.destin i.type_trip, rob
```

Figure 6.6

```
reg tot_spend los book_time i.destin i.type_trip
predict r, resid
kdensity r, normal
```

Stata-output 6.18

```
swilk r
```

The analysis in 6.18 presupposes that the variable r was made when creating Figure 6.6.

Stata-output 6.19

```
corr tot_spend los
```

Stata-output 6.20

```
oneway tot_spend destin, tab
```

Stata-output 6.21

```
esize twosample tot_spend, by(destin) cohensd hedgesg
```

Stata-output 6.22

```
tab accom destin, col chi2
```

Stata-output 6.23

```
phi accom destin
```

Before doing the phi-command (i.e., Cohen's *w*), you must first download it. In the Command-window, type findit snp3 and follow the instructions after first clicking on snp3 from http://www.stata.com/stb/stb3.

Figure 6.7

```
reg tot_spend los book_time i.destin i.type_trip
coefplot, xline(0)
```

Before doing the coefplot-command, you must first download it. In the Command-window, type findit coefplot and follow the instructions after first clicking on SJ-15-1 gr0059_1.

Figure 6.8

```
reg tot_spend los book_time i.destin i.type_trip
margins, at(los=(1 10 20 30 40 50 60))
marginsplot
```

Figure 6.9

```
cibar tot_spend, over(destin)
```

6.10.2 SPSS-Commands in Syntax-Files

Stata–output 6.14 (SPSS–output 4.1)

```
REGRESSION
  /DESCRIPTIVES MEAN STDDEV CORR SIG N
  /MISSING LISTWISE
  /STATISTICS COEFF OUTS R ANOVA
  /CRITERIA=PIN(.05) POUT(.10)
  /NOORIGIN
  /DEPENDENT tot_spend
  /METHOD=ENTER los book_time destin type_trip.
```

Figure 6.3 (length of stay)
To the best of my knowledge, there is no SPSS option readily available for generating this scatterplot smoother for all *x*-variables simultaneously. Yet there is a way to get a plot that considers one *x*-variable at a time. The commands are:

```
GGRAPH
  /GRAPHDATASET NAME="graphdataset" VARIABLES=los tot_spend MISS-
ING=LISTWISE REPORTMISSING=NO
  /GRAPHSPEC SOURCE=INLINE
  /FITLINE TOTAL=NO.
BEGIN GPL
  SOURCE: s=userSource(id("graphdataset"))
  DATA: los=col(source(s), name("los"))
  DATA: tot_spend=col(source(s), name("tot_spend"))
  GUIDE: axis(dim(1), label("los"))
  GUIDE: axis(dim(2), label("tot_spend"))
  GUIDE: text.title(label("Simple Scatter of tot_spend by los"))
  ELEMENT: point(position(los*tot_spend))
END GPL.
```

Now, right-click on the graph to get it "active." Then click on Edit Content → In Separate Window, and click on Elements and choose Fit Line at Total. Now, in the appearing Properties-Window, click on Loess and on Apply (and Close). The appearing graph resembles Figure 6.3 with length of stay at the *x*-axis.

Figure 6.3 (booking time)
To the best of my knowledge, there is no SPSS option readily available for generating this scatterplot smoother for all *x*-variables simultaneously. But there is a way to get a plot that considers one *x*-variable at a time. The commands are:

```
GGRAPH
  /GRAPHDATASET NAME="graphdataset" VARIABLES=book_time tot_spend
```

```
MISSING=LISTWISE REPORTMISSING=NO
  /GRAPHSPEC SOURCE=INLINE
  /FITLINE TOTAL=NO.
BEGIN GPL
  SOURCE: s=userSource(id("graphdataset"))
  DATA: book_time=col(source(s), name("book_time"))
  DATA: tot_spend=col(source(s), name("tot_spend"))
  GUIDE: axis(dim(1), label("book_time"))
  GUIDE: axis(dim(2), label("tot_spend"))
  GUIDE: text.title(label("Simple Scatter of tot_spend by
book_time"))
  ELEMENT: point(position(book_time*tot_spend))
END GPL.
```

Now, right-click on the graph to get it "active." Then click on Edit Content → In Separate Window, and click on Elements and choose Fit Line at Total. Now, in the appearing Properties-Window, click on Loess and on Apply (and Close). The appearing graph resembles Figure 6.3 with booking time at the *x*-axis.

Stata–output 6.15

```
REGRESSION
  /MISSING LISTWISE
  /STATISTICS COEFF OUTS R ANOVA COLLIN TOL
  /CRITERIA=PIN(.05) POUT(.10)
  /NOORIGIN
  /DEPENDENT tot_spend
  /METHOD=ENTER los book_time destin type_trip.
```

Figure 6.4

```
GGRAPH
  /GRAPHDATASET NAME="graphdataset" VARIABLES=los tot_spend MISS-
ING=LISTWISE REPORTMISSING=NO
  /GRAPHSPEC SOURCE=INLINE
  /FITLINE TOTAL=YES.
BEGIN GPL
  SOURCE: s=userSource(id("graphdataset"))
  DATA: los=col(source(s), name("los"))
  DATA: tot_spend=col(source(s), name("tot_spend"))
  GUIDE: axis(dim(1), label("los"))
  GUIDE: axis(dim(2), label("tot_spend"))
  GUIDE: text.title(label("Simple Scatter with Fit Line of tot_spend
by los"))
  ELEMENT: point(position(los*tot_spend))
END GPL.
```

Figure 6.5

```
REGRESSION
  /MISSING LISTWISE
  /STATISTICS COEFF OUTS R ANOVA COLLIN TOL
  /CRITERIA=PIN(.05) POUT(.10)
  /NOORIGIN
  /DEPENDENT tot_spend
  /METHOD=ENTER los book_time destin type_trip
  /SCATTERPLOT=(*ZRESID ,*ZPRED).
```

Stata-outputs 6.16 and 6.17

```
UNIANOVA tot_spend WITH los book_time destin type_trip
  /METHOD=SSTYPE(3)
  /INTERCEPT=INCLUDE
  /PRINT F BP
  /CRITERIA=ALPHA(.05)
  /ROBUST=HC3
  /DESIGN=los book_time destin type_trip.
```

Figure 6.6

```
REGRESSION
  /MISSING LISTWISE
  /STATISTICS COEFF OUTS R ANOVA COLLIN TOL
  /CRITERIA=PIN(.05) POUT(.10)
  /NOORIGIN
  /DEPENDENT tot_spend
  /METHOD=ENTER los book_time destin type_trip
  /SCATTERPLOT=(*ZRESID ,*ZPRED)
  /RESIDUALS HISTOGRAM(ZRESID) .
```

Stata-output 6.18

```
REGRESSION
  /MISSING LISTWISE
  /STATISTICS COEFF OUTS R ANOVA COLLIN TOL
  /CRITERIA=PIN(.05) POUT(.10)
  /NOORIGIN
  /DEPENDENT tot_spend
  /METHOD=ENTER los book_time destin type_trip
  /SCATTERPLOT=(*ZRESID ,*ZPRED)
  /RESIDUALS HISTOGRAM(ZRESID)
  /SAVE RESID.
EXAMINE VARIABLES=RES_1
  /PLOT BOXPLOT STEMLEAF NPPLOT
  /COMPARE GROUPS
  /STATISTICS DESCRIPTIVES
  /CINTERVAL 95
  /MISSING LISTWISE
  /NOTOTAL.
```

Stata-output 6.19

```
CORRELATIONS
  /VARIABLES=tot_spend los
  /PRINT=TWOTAIL NOSIG
  /MISSING=PAIRWISE.
```

Stata-output 6.20

```
ONEWAY tot_spend BY destin
  /STATISTICS DESCRIPTIVES
  /MISSING ANALYSIS.
```

Stata-output 6.21

SPSS version 27 has built-in commands to calculate Cohen's *d* or Hedges' *g*, but my version (26) has not. To compute these effect size measures semi-automatically, go to for example https://memory.psych.mun.ca/models/stats/effect_size.shtml and plug the relevant means, standard deviations, and group sizes into the "unbiased" calculator.

Stata outputs 6.22 and 6.23

```
CROSSTABS
  /TABLES=accom BY destin
  /FORMAT=AVALUE TABLES
  /STATISTICS=CHISQ PHI
  /CELLS=COUNT COLUMN
  /COUNT ROUND CELL.
```

Figure 6.7
There is no SPSS option readily available for generating this graph to the best of my knowledge.

Figure 6.8
There is no SPSS option readily available for generating this graph to the best of my knowledge.

Figure 6.9

```
GGRAPH
  /GRAPHDATASET NAME="graphdataset" VARIABLES=destin MEANCI(tot_
spend, 95) [name="MEAN_tot_spend"
    LOW="MEAN_tot_spend_LOW" HIGH="MEAN_tot_spend_HIGH"] MISS-
ING=LISTWISE REPORTMISSING=NO
  /GRAPHSPEC SOURCE=INLINE.
BEGIN GPL
  SOURCE: s=userSource(id("graphdataset"))
  DATA: destin=col(source(s), name("destin"), unit.category())
  DATA: MEAN_tot_spend=col(source(s), name("MEAN_tot_spend"))
  DATA: LOW=col(source(s), name("MEAN_tot_spend_LOW"))
  DATA: HIGH=col(source(s), name("MEAN_tot_spend_HIGH"))
  GUIDE: axis(dim(1), label("destin"))
```

```
GUIDE: axis(dim(2), label("Mean tot_spend"))
GUIDE: text.title(label("Simple Bar Mean of tot_spend by destin"))
GUIDE: text.footnote(label("Error Bars: 95% CI"))
SCALE: linear(dim(2), include(0))
ELEMENT: interval(position(destin*MEAN_tot_spend), shape.interi-
or(shape.square))
ELEMENT: interval(position(region.spread.range(des-
tin*(LOW+HIGH))), shape.interior(shape.ibeam))
END GPL.
```

6.11 Chapter Exercises with Solutions

The exercises below use the data available for download on the book's website.

Exercises:

Exercise 1 (data: `student_fem_weight`, see appendix A of this chapter for data documentation)

    1a Find the formula for Body Mass Index (BMI) on the Internet and compute this new variable based on the relevant variables in the data.

    1b What is the mean of the BMI variable computed in 1a? The minimum? The maximum? The range? The median? The SD? The CV?

    1c Why are there only 289 valid units for the BMI variable?

    1d Is there an association between age and BMI? If so, how strong is this association?

    1e Is there a statistically significant difference in BMI between students who never snuff and those who do? If so, how large is this difference?

    1f Is the association between `weight_kg` and `height_cm` linear? Are there any outliers in this association?

    1g Is there an association between `qol` and `health`? If so, how strong is this association?

Exercise 2 (data: `red_wine`, see appendix C of Chapter 4 for data documentation)

    2a Estimate the following regression model: `log_price` by the x-variables `quality`, `age` and `in_store`. Describe your results.

    2b For the model in 2a: Are the effects of `quality` and `age` linear?

    2c For the model in 2a: Is there a problem with multicollinearity?

    2d Using the model in 2a as point of departure, is there an interaction effect between `in_store` and `quality`? What does this mean in terms of the additivity assumption?

    2e For the model in 2d: Examine the following assumptions: homoscedasticity in error terms (C1), normally distributed error terms (C2), and uncorrelated error terms (C3).

Answers to exercises (mainly in Stata; see Section 6.10 for equivalent SPSS syntaxes):

Exercise 1 (data: `student_fem_weight`, see appendix A of this chapter for data documentation)

    1a Find the formula for Body Mass Index (BMI) on the Internet and compute this new variable based on the relevant variables in the data.

There are two main formulas to compute BMI[36]:

(1) weight in kg / (height in meters)^2

or

(2) (weight in kg / (height in cm)^2) × 10,000

Since the height variable in the data refers to cm, we should use (2) to compute the BMI variable.[37] In Stata, the (do-file) command becomes:

```
gen bmi = (weight_kg/(height_cm*height_cm))*10000
```

In SPSS, the analogous syntax is:

```
COMPUTE bmi=(weight_kg/(height_cm*height_cm))*10000.
EXECUTE.
```

1b What is the mean of the BMI variable computed in 1a? The minimum? The maximum? The range? The median? The SD? The CV?

We could apply the sum and tab commands, but again it is faster to use the tabstat-command from Section 2.7 as in:

```
. tabstat bmi, stats(count mean min max range median sd cv)

    variable |         N        mean         min         max       range         p50
-------------+------------------------------------------------------------------------
         bmi |       289    23.42376     15.0597    37.61841    22.55871    23.03004
------------------------------------------------------------------------------------

    variable |        sd          cv
-------------+-----------------------
         bmi |  3.326076    .1419958
-----------------------------------
```

The average student has a BMI of about 23. (A BMI in the range from 18.5 to 23.9 is considered as "healthy" according to one rule-of-thumb.) BMI's range is 23 points, ranging from about 15 to about 38. The median is very similar to the mean, making the BMI-distribution symmetrical. The SD and CV are 3.33 and 0.14, respectively.

1c Why are there only 289 valid units for the BMI variable?

Only 289 out of the 331 female students in the data answered the question on their weight. It is thus only possible to compute the BMI for these 289 students.

1d Is there an association between age and BMI? If so, how strong is this association?

```
. reg bmi age

      Source |       SS           df       MS            Number of obs   =       289
-------------+------------------------------          F(1, 287)       =     11.21
       Model |  119.781856         1   119.781856       Prob > F        =    0.0009
    Residual |  3066.30006       287   10.6839723       R-squared       =    0.0376
-------------+------------------------------          Adj R-squared   =    0.0342
       Total |  3186.08191       288   11.0627844       Root MSE        =    3.2686
```

```
----------------------------------------------------------------------------
        bmi |      Coef.   Std. Err.        t    P>|t|     [95% Conf. Interval
------------+---------------------------------------------------------------
        age |    .1179966   .0352404      3.35   0.001     .0486343    .187358
       _cons |   20.68126    .841329     24.58   0.000     19.0253    22.3372
----------------------------------------------------------------------------
```

```
. corr bmi age
(obs=289)

             |      bmi      age
-------------+------------------
        bmi  |   1.0000
        age  |   0.1939   1.0000
```

Yes, there is a positive and statistically significant association ($p = 0.001$) between the two variables: Older female students on average appear to have greater BMIs than younger female students. The association is "small" ($r = 0.19$) according to the guidelines mentioned in Section 6.7. Yet we should always remember that such effect size classifications are rules-of-thumb. Ideally, assessment of effect sizes should be judged against the results in prior research.

1e   Is there a statistically significant difference in BMI between students who never snuff and those who do? If so, how large is this difference?

```
. reg bmi i.snuff

      Source |       SS           df       MS        Number of obs   =       289
-------------+----------------------------------     F(1, 287)       =      0.45
       Model |  4.93648252         1   4.93648252    Prob > F        =    0.5051
    Residual |  3181.14543       287   11.0841304    R-squared       =    0.0015
-------------+----------------------------------     Adj R-squared   =   -0.0019
       Total |  3186.08191       288   11.0627844    Root MSE        =    3.3293

----------------------------------------------------------------------------
        bmi |      Coef.   Std. Err.        t    P>|t|     [95% Conf. Interval
------------+---------------------------------------------------------------
      snuff |
sometimes/daily |  .2822344   .4229139      0.67   0.505    -.5501718   1.11464
       _cons |   23.33587   .2360067     98.88   0.000     22.87134   23.8003
----------------------------------------------------------------------------
```

Female students who snuff have a 0.28 points greater BMI on average than female students who do not. But since the *p*-value is 0.51, we cannot reject the null hypothesis that the two groups have similar BMIs in the population.[38] Since the answer to the first question is no, the second question becomes more or less redundant.

1f Is the association between `weight_kg` and `height_cm` linear? Are there any outliers in this association?

Using a scatterplot smoother (i.e., the `mrunning`-command in Stata), we may in this case answer both questions at the same time. The plot shows an approximately linear association with only a few potentially influential outliers.

1g Is there an association between `qol` and `health`? If so, how strong is this association?

I assume that quality of life (`qol`) in the main is determined by physical health level (`health`), but I concede that the association might go the other way round. We have:

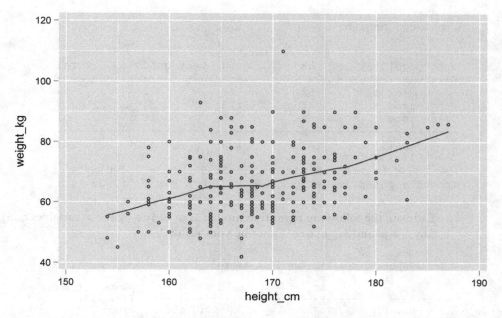

Figure 6.10 Scatterplot smoother visualizing the association between the variables weight_
kg and height_cm.

```
. tab qol health, col chi2

              |             health
         qol |         ok       good  very good |      Total
-------------+---------------------------------+----------
          ok |         62         28          2 |         92
             |      44.60      17.61       6.06 |      27.79
-------------+---------------------------------+----------
        good |         71        116         21 |        208
             |      51.08      72.96      63.64 |      62.84
-------------+---------------------------------+----------
   very good |          6         15         10 |         31
             |       4.32       9.43      30.30 |       9.37
-------------+---------------------------------+----------
       Total |        139        159         33 |        331
             |     100.00     100.00     100.00 |     100.00

       Pearson chi2(4) =  50.5559   Pr = 0.000
```

We note the unexpected pattern that having good or very good health appears to
be "beneficial" for experiencing better quality of life (*p* < 0.0001). The answer to
the first question is yes. The answer to the second question is a medium-sized effect
according to the rules-of-thumb mentioned in Section 6.7; cf. the table showing a
Cohen's *w* of 0.39:
Exercise 2 (data: red_wine, see appendix C of Chapter 4 for data documentation)

```
. phi qol health

              |              health
        qol |       ok       good  very good |       Total
------------+-------------------------------+----------
         ok |       62         28         2 |          92
       good |       71        116        21 |         208
  very good |        6         15        10 |          31
------------+-------------------------------+----------
      Total |      139        159        33 |         331

            Pearson chi2(4) =    50.5559    Pr = 0.000
   Cramer's phi-prime =   0.2763       Cohen's w = 0.3908
```

2a   Estimate the following regression model: log_price by the x-variables qual-
     ity, age and in_store. Describe your results.

```
. reg log_price quality age i.in_store

      Source |        SS          df       MS          Number of obs   =        21
-------------+----------------------------------       F(3, 214)       =     103.0
       Model |   62.3267599        3   20.7755866       Prob > F        =     0.000
    Residual |   43.1506897      214   .201638737       R-squared       =     0.590
-------------+----------------------------------       Adj R-squared   =     0.585
       Total |   105.47745       217   .486071196       Root MSE        =     .4490


-------------------------------------------------------------------------------
   log_price |     Coef.    Std. Err.       t     P>|t|     [95% Conf. Interval]
-------------+-----------------------------------------------------------------
     quality |   .0618344    .0053191    11.63    0.000      .05135       .072318
         age |   .1391664    .0152362     9.13    0.000      .1091341     .169198
             |
    in_store |
         Yes |  -.2194535    .0748975    -2.93    0.004     -.3670847    -.071822
       _cons |  -2.465478    .4443145    -5.55    0.000     -3.341271    -1.58968
-------------------------------------------------------------------------------
```

A quick summary of the results assuming the data comprise a random sample from a
well-defined population is this: Better tasting and older wines are more expensive on average than worse tasting and younger wines *ceteris paribus*. A one-point increase in quality
suggests a 6.2 percent more expensive wine; a one-year increase in storage suggests a 13.9
percent pricier wine – allowing in part for two dynamic interpretations. Wines available
in store on average cost about 20 percent less than wines that need to be ordered. We apply the formula $100 \times (e^b - 1)$ from Section 4.5 to get this percentage difference. That is,
$100 \times (2.7182^{-.2194535} - 1) \approx -19.70.$[39] All regression coefficients are statistically significant
(at $p < 0.005$), and R^2 suggests that the three x-variables combined account for 59 percent
of the variation (variance) in wine prices. Whether this is a satisfactory model fit or not
should be evaluated against some yardstick – typically found in prior research.

   2b   For the model in 2a: Are the effects of quality and age linear?
   The plots suggest that both effects are roughly linear, making the linear regression
model suitable at the outset.[40] That said, the effect of taste quality appears to have something of a breaking point at about 82 points.

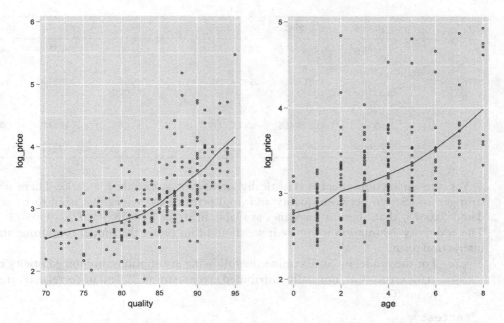

Figure 6.11 Multivariable scatterplot smoother visualizing the associations between `log_price` and `quality` and `log_price` and `age`.

2c   For the model in 2a: Is there a problem with multicollinearity?

```
. vif

    Variable |       VIF       1/VIF
-------------+----------------------
     quality |      1.08    0.929893
         age |      1.06    0.946169
  1.in_store |      1.03    0.974995
-------------+----------------------
    Mean VIF |      1.05
```

No, there is no problem with multicollinearity. The mean VIF is much lower than any dangerous threshold level.

2d   Using the model in 2a as a point of departure, is there an interaction effect between `in_store` and `quality`? What does this mean in terms of the additivity assumption?

```
. reg log_price age i.in_store##c.quality

      Source |       SS           df       MS      Number of obs   =       218
-------------+----------------------------------   F(4, 213)       =     79.28
       Model |  63.0973738         4  15.7743434   Prob > F        =    0.0000
    Residual |  42.3800758       213  .198967492   R-squared       =    0.5982
-------------+----------------------------------   Adj R-squared   =    0.5907
       Total |   105.47745       217  .486071196   Root MSE        =    .44606
```

log_price	Coef.	Std. Err.	t	P>|t|	[95% Conf.	Interval]
age	.1386208	.0151375	9.16	0.000	.1087824	.168459
in_store						
Yes	1.708906	.9826728	1.74	0.083	-.2281032	3.64591
quality	.0678729	.00611	11.11	0.000	.0558291	.079916
in_store#c.quality						
Yes	-.0231655	.011771	-1.97	0.050	-.0463681	.000037
_cons	-2.976126	.5119834	-5.81	0.000	-3.985329	-1.96692

Yes, the interaction effect is statistically significant at exactly $p = 0.05$. There is a stronger association between quality and (log) price of the wines that need to be ordered (0.0679) than there is for wines available in stores (0.0678 − 0.0232 = 0.0446). The additivity assumption is violated; we should thus prefer the model containing the interaction term.

   2e   For the model in 2d: Examine the following assumptions: homoscedasticity in error terms (C1), normally distributed error terms (C2), and uncorrelated error terms (C3).

```
. hettest
Breusch-Pagan / Cook-Weisberg test for heteroskedasticity
        Ho: Constant variance
        Variables: fitted values of log_price

        chi2(1)      =      28.72
        Prob > chi2  =     0.0000
```

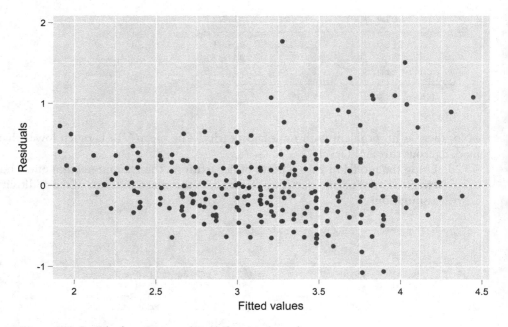

Figure 6.12 Residuals–versus–predicted/fitted-values plot.

Both test and plot suggest heteroscedasticity, that is, a violation the homoscedasticity assumption.

The remedy in the face of heteroscedasticity is to compute robust SEs:

```
. reg log_price age i.in_store##c.quality, r

Linear regression                              Number of obs   =         218
                                               F(4, 213)       =       76.91
                                               Prob > F        =      0.0000
                                               R-squared       =      0.5982
                                               Root MSE        =      .44606

------------------------------------------------------------------------------
                  |               Robust
       log_price |      Coef.   Std. Err.      t    P>|t|     [95% Conf. Interval]
------------------+-----------------------------------------------------------
             age |    .1386208    .017022     8.14   0.000     .1050677    .1721739
                 |
        in_store |
             Yes |    1.708906   .7279488     2.35   0.020     .2739995    3.143812
         quality |    .0678729    .006428    10.56   0.000     .0552023    .0805435
                 |
in_store#c.quality |
             Yes |   -.0231655   .0088551    -2.62   0.010    -.0406204   -.0057106
                 |
           _cons |   -2.976126   .5325854    -5.59   0.000    -4.025939   -1.926313
------------------------------------------------------------------------------
```

The results of the robust regression suggest no far-reaching consequences for the *p*-values.

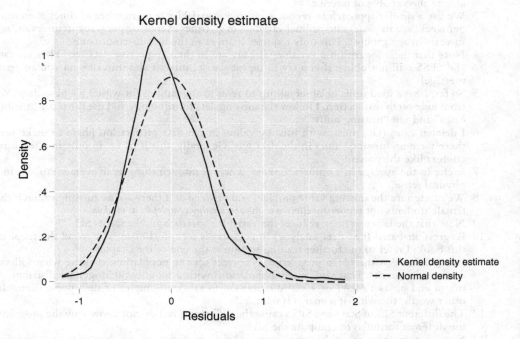

Figure 6.13 Kernel density plot and the normal distribution for the residuals.

```
. swilk r

             Shapiro-Wilk W test for normal data

    Variable |      Obs        W        V        z      Prob>z
-------------+-------------------------------------------------
           r |      218   0.95652    6.993    4.494    0.00000
```

Both plot and test indicate non-normally distributed error terms, that is, a violation of the normality assumption. Yet this has no consequences of substance in our setting given the large number of units in the analysis.

We have uncorrelated errors if the price of one wine in general is unassociated with the price of another wine. This could be the case in real life, but it is also thinkable that wines belonging to the same district tend to "share" the price level of that district irrespective of differences in quality or storage. In the latter case, we cannot rule out uncorrelated errors completely. We need more information regarding the data collection phase to answer this question properly.

Notes

1 A second reason for collapsing is to avoid having categories with very few units. More categories inevitably mean more categories with fewer units given a fixed number of units in the data.
2 The third line creates the labels for a temporary help-variable that I personally always name with an «l» at the end (short for label). The fourth line attaches the labels for the help-variable to the variable of interest.
3 We use a similar approach to recode continuous variables (e.g., number of times exercising per week, age in years) into ordinal variables (e.g., times exercising per week in intervals, age in years in age groups). This only requires changes in the recode-commands.
4 Note that these logarithmic variables already are in their respective data sets. Hence, Stata and SPSS will not oblige if you try to repeat the commands and instead send you an error message!
5 So far, I have used units or observations to refer to the entities for which we have data. Yet from here on in this section, I follow the missing data terminology and use the term "missing cases" and not "missing units."
6 I deleted cases (i.e., units) with missing values in the data preparation phase to make sure there were no missing values in the data sets. Generally, this is ok to do only in a teaching context like the present.
7 I refer to the average in a numerical sense. There is no such thing as an average student in a physical sense!
8 We can ignore the missing value problem and proceed as if there are no missing values if the female students not reporting their weight are *missing completely at random*.
9 Note that the latter strategy reduces the variable's variance and hence its SD.
10 Regression-based (i.e., data-analytic) imputation is technical material beyond the scope of this book; I refer to the further reading section at the end of the chapter.
11 I emphasize that the 65:35 proportion is correct for the population, making what follows a what-if scenario. That said, many populations with a roughly 50:50 gender distribution might end up as a 60:40 or 40:60 distribution in random samples of such populations. In other words, the what-if scenario is relevant!
12 The different SEs of Stata and SPSS cause the different CIs. I do not know why the programs use different formulas to compute the SE.
13 Note that in order to apply sampling weights to background variables (e.g., age, gender, county of residence, educational level etc.) we must know beforehand the correct population

distribution for these variables. Such information is not always available. Typically, we use sampling weights for several background variables simultaneously.

14 It is debatable whether a statistical significance assessment is relevant for data like these, as discussed in Section 5.8. Yet I choose to follow convention and make the assessment in a typical manner.

15 There are several ways of restricting an analysis to a subgroup of units in Stata. Here, I use `if homi < 4` because I know that all countries save for the USA have a lower homicide rate than four.

16 I am not claiming that the USA *should* be discarded from the data; the point is to illustrate what outliers (might) do.

17 I use the term "unbiased" and not causal in this section to avoid getting into more complexities than necessary. I also assume we analyze a random sample from some well-defined population.

18 The non-testability of A1 is the main reason why experimental control trumps statistical control when it comes to identifying causal effects with (more) certainty.

19 For example, it is sometimes interesting to find out that an x-variable that according to theory and prior research should have a significant effect indeed has not.

20 Yet using a scatterplot smoother (e.g., the `mrunning`-command in Stata) to check the linearity assumption might also get you a long way in terms of detecting outliers.

21 The results of exploratory work suggest no outlier problem for the association between length of stay and total trip expenditures.

22 Technically, only perfect multicollinearity is a violation of B4. Yet near-perfect multicollinearity also causes trouble. The latter is what I explore here.

23 Conversely, it would be difficult for a regression to find out if the small total expenditures of a particular trip were the result of being (1) a short-lasting trip or being (2) a Nordic destination trip.

24 Inflated SEs are another symptom of multicollinearity: All else being equal, it gets more difficult to obtain a statistically significant regression coefficient in the presence of multicollinearity.

25 I repeat: The same applies to b_2, b_3, and so on.

26 Note also that logging the y-variable, cf. Section 4.4, in many applications, will mitigate heteroscedasticity or make the problem totally disappear.

27 Out of 644 predictions, the LPM in Stata-output 4.18 generated four predictions below zero and none above one. This is hardly problematic.

28 Cohen's w also equals Cramér's V. Yet neither measure is restricted to two dummies; all strictly categorical may be used. For effect sizes regarding associations involving two ordinal variables, I again refer to Agresti (2010).

29 In the two-group setting, as in the present case, the F-test and the t-test are equivalent.

30 No asterisk implies $p > 0.05$ or not significant at the 5-percent level. If a regression coefficient (or any other statistical association) refers to a small sample, say 100 units or less, one asterisk might indicate the 7-percent significance level, two asterisks the 5-percent level, and three asterisks the 1-percent level.

31 Note that overlapping CIs for two groups does *not* necessarily imply a non-significant mean group difference.

32 Always use fewer PPs than you initially prepared, and always put fewer words/tables/graphs on each PP than you think is proper: Less is more! Do not read out aloud what the PPs say; the audience knows how to read!

33 Beware of the danger of overestimating the audience's numerical ability, however. This overestimating tendency is in my experience more prevalent than underestimating it.

34 I use the singular y for ease of presentation only.

35 If you have observational data, that is. If your data are experimental, dynamic interpretations might be justified.

36 See, for example, https://www.cdc.gov/nccdphp/dnpao/growthcharts/training/bmiage/page5_1.html.

37 We could alternatively rescale the height variable to meters and use (1).

38 Presupposing a non-directional H_1. A one-way ANOVA of course yields the same result.

39 In Stata, the (do-file) command is: dis 100 × (exp(−0.2194535)−1)

40 The command in Stata is: `mrunning log_price quality age`

Appendix A
Female Student Weight Data

Data documentation for the data `student_fem_weight`; a survey questionnaire data from a random sample of female students attending a Norwegian university college in 2018. Variable names are in **bold** typeface. $N = 289 - 331$.

health

Your physical health level: ok = 0, good = 1, very good = 2

qol

Your quality of life: ok = 0, good = 1, very good = 2

youth_exe

In your youth before you started studying, to what extent were you involved in sports that required a lot of physical exercising (to a very small extent = 1; to a very great extent = 10)?

age

Age in years

snuff

Snuffing (moist snuffing): never = 0, sometimes/daily = 1

weight_kg

Weight in kilograms (kg)

height_cm

Height in centimeters (cm)

References

Aggerwal, C. C. (2017). *Outlier Analysis*. New York: Springer International Publishing.

Agresti, A. (2010). *Analysis of Ordinal Categorical Data*. Second Edition. Hoboken, NJ: Wiley.

Agresti, A. (2018). *Statistical Methods for the Social Sciences*. Fifth Edition. London, UK: Pearson Education Limited.

Allison, P. D. (1999). *Multiple Regression. A Primer*. Thousand Oaks, CA: Pine Forge Press.

Allison, P. D. (2002). *Missing Data*. Newbury Park, CA: Sage Publications, Inc.

Angrist, J. D. and Pischke, J.-S. (2009). *Mostly Harmless Econometrics. Am Empiricist's Guide*. Princeton, NJ: Princeton University Press.

Angrist, J. D. and Pischke, J.-S. (2015). *Mastering 'Metrics. The Path from Cause to Effect*. Princeton, NJ: Princeton University Press.

Berk, R. A. (2004). *Regression Analysis. A Constructive Critique*. Thousand Oaks, CA: Sage Publications, Inc.

Berry, W. D. (1993). *Understanding Regression Assumptions*. Newbury Park, CA: Sage Publications, Inc.

Best, H. and Wolf, C. (2015). *The Sage Handbook of Regression Analysis and Causal Inference*. London, UK: Sage Publications, Inc.

Bittmann, F. (2019). *STATA. A Really Short Introduction*. Berlin, Germany: De Gruyter Oldenbourg.

Bryman, A. (2016). *Social Research Methods*. Fifth Edition. Oxford, UK: Oxford University Press.

Cohen, J. (1988). *Statistical Power Analysis for the Behavioral Sciences*. Second Edition. Hillsdale, NJ: Lawrence Erlbaum Associates, Publishers.

Cunningham, S. (2021). *Causal Inference: The Mixtape*. London, UK: Yale University Press.

Daniels, L. and Minot, N. (2020). *An Introduction to Statistics and Data Analysis Using Stata*. London, UK: Sage Publications, Inc.

Field, A. (2018). *Discovering Statistics Using IBM SPSS Statistics*. London, UK: Sage Publications, Inc.

Freedman, D., Pisani, R. and Purves, R. (2007). *Statistics*. Fourth Edition. New York, NY: W.W. Norton & Company.

Frick, R. W. (1998). Interpreting Statistical Testing: Process and Propensity, Not Population and Random Sampling. *Behavior Research Methods, Instruments, & Computers*, 30, 527–535.

Gelman, A. and Nolan, D. (2017). *Teaching Statistics. A Bag of Tricks*. Second Edition. Oxford, UK: Oxford University Press.

Gorard, S. (2019). Do We Really Need Confidence Intervals in the New Statistics? *International Journal of Social Science Methodology*, 22, 281–291.

Gorard, S. (2020). Handling Missing Data in Numerical Analyses. *International Journal of Social Science Methodology*, 23, 651–660.

Healey, K. (2019). *Data Visualization. A Practical Introduction*. Princeton, NJ: Princeton University Press.

Hosmer, D. W., Lemeshow, S. and Sturdivant, R. X. (2013). *Applied Logistic Regression*. Third Edition. Hoboken, NJ: John Wiley & Sons.

Imbens, G. W. and Rubin, D. B. (2015). *Causal Inference for Statistics, Social, and Biomedical Sciences. An Introduction*. New York, NY: Cambridge University Press.

Kahneman, D. (2011). *Thinking, Fast and Slow*. New York, NY: Farrar, Straus and Giroux.

Kennedy, P. (2002). Sinning in the Basement: What Are the Rules? The Ten Commandments of Applied Econometrics. *Journal of Economic Surveys*, 16, 569–589.

Khamis, H. (2008). Measures of Association. How to Choose? *Journal of Diagnostic Medical Sonography*, 24, 155–162.

Kline, R. B. (2020). *Becoming a Behavioral Science Researcher. A Guide to Producing Research That Matters*. Second Edition. New York, NY: The Guilford Press.

Lew, M. J. (2012). Bad Statistical Practice in Pharmacology (and other Basic Biomedical Disciplines): You Probably Don't Know *P*. *British Journal of Pharmacology*, 166, 1559–1567.

Long, J. S. (1997). *Regression Models for Categorical and Limited Dependent Variables*. Thousand Oaks, CA: Sage Publications, Inc.

Long, J. S. (2015). Regression Models for Nominal and Ordinal Outcomes. In Best, H. and C. Wolf: *The Sage Handbook of Regression Analysis and Causal Inference*, pp. 173–203. London, UK: Sage Publications, Inc.

Meuleman, B., Loosveldt, G. and Emonds, V. (2015). Regression Analysis: Assumptions and Diagnostics. In Best, H. and C. Wolf: *The Sage Handbook of Regression Analysis and Causal Inference*, pp. 83–110. London, UK: Sage Publications, Inc.

Mitchell, M. N. (2021). *Interpreting and Visualizing Regression models Using Stata*. Second Edition. Lakeway Drive, College Station, TX: Stata Press.

Pearl, J. and Mackenzie, D. (2018). *The Book of Why. The New Science of Cause and Effect*. New York, NY: Basic Books.

Reinhart, A. (2015). *Statistics Done Wrong*. San Francisco, CA: No Starch Press, Inc.

Rictchie, S. (2020). *Science Fictions. How Fraud, Bias, Negligence, and Hype Undermine the Search for Truth*. New York, NY: Metropolitan Books.

Rosenbaum, P. R. (2017). *Observation & Experiment. An Introduction to Causal Inference*. Cambridge, MA: Harvard University Press.

Salsburg, D. (2001). *The Lady Tasting Tea. How Statistics Revolutionized Science in the Twentieth Century*. New York, NY. Holt Paperbacks.

Schneider, J. W. (2013). Caveats for Using Statistical Significance Tests in Research Assessments. *Journal of Informetrics*, 7, 50–62.

Sirkin, R. M. (2005). *Statistics for the Social Sciences*. Third edition. London, UK: Sage Publications, Inc.

Spiegelhalter, D. (2019). *The Art of Statistics – Learning from Data*. London, UK: Penguin Books.

Stigler, S. M. (2016). *The Seven Pillars of Statistical Wisdom*. Cambridge, MA. Harvard University Press.

Thrane, C. (2020). *Applied Regression Analysis. Doing, Interpreting and Reporting*. London, UK: Routledge.

Vries, R. de. (2019). *Critical Statistics. Seeing beyond the Headlines*. London, UK: Red Globe Press.

Weisberg, H. I. (2014). *Willful Ignorance. The Mismeasure of Uncertainty*. Hoboken, NJ: Wiley.

Wheelan, C. (2013). *Naked Statistics*. New York, NY: W.W. Norton & Company.

White, P. and Gorard, S. (2017). Against Inferential statistics: How and Why Current Statistics Teaching Gets It Wrong. *Statistics Education Research Journal*, 16, 55–65.

Wolf, C. and Best, H. (2015). Linear Regression. In Best, H. and C. Wolf: *The Sage Handbook of Regression Analysis and Causal Inference*, pp. 55–81. London, UK: Sage Publications, Inc.

Wooldridge, J. M. (2010). *Econometric Analysis of Cross Section and Panel Data*. Cambridge, MA: The MIT Press.

Index

Note: *Italic* page numbers refer to figures and page numbers followed by "n" denote endnotes.

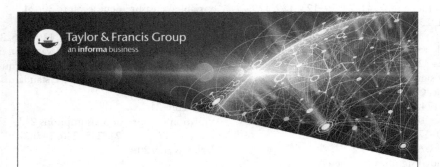

Printed in the United States
by Baker & Taylor Publisher Services

Printed in the United States
by Baker & Taylor Publisher Services